THE PATENT JUNGLE

THE PATENT JUNGLE

The Inventor's Friendly User Guide

**(includes guidelines to follow under latest changes
in the 2011 patent law—"America Invents")**

Leon Cooper

Published by 90 Day Wonder Publishing
P. O. Box 6030, Malibu, CA 90265
leoncooper@verizon.net

Printed in the United States of America

ISBN 13: 978-0-9790584-7-9
ISBN 10: 0-9790584-7-3

Price: $10.00
For quantity discounts
contact leoncooper@verizon.net.

"Human subtlety will never devise an invention more beautiful, more simple or more direct than does nature because in her inventions nothing is lacking, and nothing is superfluous."
– Leonardo da Vinci

Contents

INTRODUCTION

I've written this booklet because I've "been there"—with an idea for a product I was convinced had merit; it would save lives, was needed, and would probably be bought by many in a large, diversified market. A good patent for this product would become the foundation for a corporation I had been planning to set up anyway.

In the 1970's, a major campaign was underway, nationwide, urging homeowners to buy and install smoke detectors. Cities were passing laws making smoke detectors mandatory in new construction; some were making the installation of smoke detectors mandatory in remodeling as well. At the time, a few manufacturers made smoke detectors equipped with "test buttons," obviously to assure the detector's owner that his safety product would give the alarm when smoke was present. Other manufacturers recommended using actual smoke from cigarettes, cigars or burning paper to test their devices. The user manuals of other manufacturers were silent on the subject of testing.

By actually testing a number of smoke detectors, I concluded that test buttons only tested the circuitry, but didn't signal the presence of smoke— didn't test the sensing mechanism. Cigarettes and cigars were unhealthful; burning paper or other combustible material was dangerous and messy. A flame too close to the smoke detector would destroy it, or at least impair its effectiveness. In any event, a simple external test was needed to signal the presence of smoke to cause the smoke detector to go into alarm.

I had studied chemistry in school, so at least I knew how to avoid hurting myself, or else blowing the place up. Referring to my chemistry handbook, I began my search for a chemical formulation that simulated smoke and would also cause both types of smoke detectors—photoelectric and ionization—to go into alarm. The formulation would be packaged in an aerosol can. That way the spray could be directed at the smoke detector vents from several feet away—no need to risk the danger of falling by standing on a chair or other elevation in order to test the smoke detector.

After many trials, Eureka! I found it, only to be told by a chemist friend of mine that the formulation was highly toxic. But I knew I was on the right track. It was only a question of more trials before I finally came up with my winning formula—one that was safe and easy to use and would cause both types of smoke detectors to go into alarm. I decided to call it what it was—Smoke Detector Tester—nothing fancy or cute. (See Appendix D for Cooper patent).

I felt I could go forward now, in confidence.

I decided to pay for a market survey, including focus groups. The results seemed to be conclusive: there was a definite need for an easy-to-use, economical product that would provide the smoke detector owner with the assurance that his safety product would give the alarm when smoke was present.

Now the really hard work began. There is a natural tendency called "inertia" among us humans to resist, to dismiss, to oppose anything that's new or different. I needed to contact the major interest groups to overcome this problem. My first contact, of course, was with the National Fire Protection Association, the organization that had started the national campaign for smoke detectors. I finally was able to meet with the Association's "code" committee, the one charged with establishing standards for fire alarms. At their meeting in Las Vegas I demonstrated my product. "It's only a functional test—only a "go no go" test—either the alarm will go off or it won't," I assured the committee members, who watched as I sprayed the two types of smoke detectors—first one then the other. Both detectors went into alarm a few seconds after I sprayed. (You may be sure it was an anxious few seconds for me). "We'll let you know, Mr. Cooper," the chairman said as I left the room.

I called Underwriters Lab in Chicago. I knew that UL approval would be contingent upon favorable action by the National Fire Protection Association. "Yes, I know," I told the UL representative, " NFPA has the matter under consideration right now…I'll let you know of their decision as soon as I find out." "Sure, it's OK to send a sample," he replied to my question.

I made other contacts in the meanwhile, giving samples to city fire prevention officers, fire city building inspectors, fire alarm installers, managers of hotel/motel chains and others. If my product worked as I had claimed, "Would they endorse it—at least say it's OK?" Most were enthusiastic; they wanted to tell others. No, I didn't "bribe" anybody. I sent a sample to Consumer Reports, also to the U.S. Consumer Products Safety Commission. I went to Seattle to talk to the Boeing. They have fire alarm systems on all of their planes, don't they? What about using my Smoke Detector Tester to test them? After demonstrating my product, I heard, "We'll let you know, Mr. Cooper." (again).

I continued making the rounds. Only by continuing contacts—as with most endeavors like this—the "carborundum principle" finally paid off. By gaining more believers, by at least dispelling doubters, my Smoke Detector Tester became known and accepted. I was always careful of being too "pushy." I didn't want anyone to say, "What? You again?"

Then, one day, "It" happened. The National Fire Protection Association committee finally returned my phone calls. "Yes, Leon (my first name now, no longer formal "Mr. Cooper"), we've approved your Smoke

Detector Tester to be used as a 'functional'—my terminology— test of fire alarms—annually…Yes, you can let Underwriters know. We'll be sending them our new code language, tell them."

One after another, my contacts began to fall into line… Underwriters Labs gave me the "green light," citing NFPA. U.S. Consumer Products Safety Commission said it was "safe" to use; Consumer Reports, with their millions of members, said, "this novel product offered an easy, convenient way of testing smoke detectors." Boeing's "Ground Support Manual" identified my product as "required" in the testing of all fire alarms aboard all of their commercial planes. Airbus followed suit soon afterwards, naturally. My doubters among a few smoke detector manufacturers withdrew their questions.

Feeling flush with all these approvals and endorsements, I decided to apply for a patent, and was prepared to spend money—well, maybe, not too much—in order to go after it. How could anyone possibly refuse to grant me a patent for my great new product? Immediately, too. Little did I anticipate the hurdles I would have to overcome in dealing with a dubious patent examiner.

There were other doubters, still—my wife (it was her money, too) , my relatives, yes, of course, my know-it-all brother-in-law, a few of my friends, others. But I persevered. I wasn't on a mission to save mankind— nothing that earth-shaking. I simply wanted to offer a product that met a need, that many would buy, that could be patented.

If this is you, and you're prepared to go ahead to spend time and money on your idea, this booklet is for you.

CHAPTER ONE

PATENTS AND OTHER INTELLECTUAL PROPERTIES

"A patent, or invention, is any assemblage of technologies
or ideas that you can put together that nobody
put together that way before."
– Anon.

The Founding Fathers deemed inventions, innovations, and new ideas to be so important for the young nation to thrive and prosper that they enshrined patent law in our country's organic act, the Constitution, right at the beginning, in Article One—no doubt at the urging of Ben Franklin, himself a renowned inventor, and by Alexander Hamilton, the businessman mind among that extraordinary group. "The Congress shall have the power ... to promote the Progress of Science and useful Arts, by securing for limited Times to Authors and Inventors exclusive Right to their respective Writings and Discoveries."

Patents are "Intellectual Properties." Patent owners and others interested in patents need to have a good understanding of other, closely related Intellectual Properties. Here are several:

"Trademark;" "Servicemark" –often used interchangeably—is a word, name, symbol, or device used in trade with goods to indicate the source of the goods and to distinguish them from the goods of others (IBM,

GM, Coca Cola, A T &T, WAL-MART).

"Copyright." A Copyright is a form of protection provided to authors of "original works of authorship," including literary, dramatic, musical, artistic, and certain other intellectual works, both published and unpublished. Ownership of a Copyright grants the exclusive right to reproduce the copyrighted work, to prepare derivative works, to distribute copies of the copyrighted work, to perform the copyrighted work publicly, or to display the copyrighted work publicly.

Legally, you own any of the foregoing—simply by using them in the course of doing business. It is best, however, to actually register with the Patent and Trademark Office (Office) for these rights, dispelling any question regarding ownership. To register for either the Trademark or the Servicemark the direct link to the electronic form is (teas/teasplus. htm). The fee for each mark is $275.00.

Of the three rights, the Copyright (administered by the Library of Congress) is the most important one to formally register if you're going to apply for a patent. You'll benefit by having it even if you don't get a patent. The best way to file for a Copyright is to do so electronically. The direct link is eCO. For $35.00 your application—best by using your credit card—will be processed more quickly, with the added advantage of tracking it. When the Library receives a physical document or CD of your Copyright you own it for the rest of your life plus 70 years. Now you have clear ownership, including a description of your product. This will stand you in good stead, proving "total" ownership of your patented invention should you have to deal with an infringer or others.

Besides, a Copyright, like a Trademark or Servicemark has monetary value and can be assigned, licensed or franchised for money while still retaining ownership. Next is another important Intellectual Property: "Trade Secret." Unlike a patent, a Trade Secret is not disclosed, is not known to the public, and the owner of a Trade Secret has taken efforts to keep it secret. Chemical formulas, e.g., Celebrex, Tylenol, are among the first examples that come to mind. But there are other Trade Secrets. If you are a business owner you shouldn't overlook other important information about your business you need to safeguard, including customer lists, accounting records, sales analyses, manufacturing processes, office procedures, "know-how," employee records, etc., in short, all of the information your business uses as a going concern .

Most states have adopted the Uniform Trade Secrets Act, so you should be familiar with the various provisions in your particular state law, especially how to avail yourself of its protections. In order to enforce your rights under these laws, you must take formal steps, including well-defined written procedures. Without them, action you might take against the offending party is significantly weakened; and you may not be able to enforce payment from someone – especially an ex-employee or a competitor—who has taken secret information about your business without your express permission. Thus anyone who has privileged information about your business must have acknowledged, in writing, that he understands this. It is important to review the status of those who are "privileged" from time to time, withdrawing some, adding others. You must be able to prove compliance with your company's policies and procedures.

Trade Secrets are not limited to a time period or to the narrowly defined principles governing patent laws and the other Intellectual Properties discussed here. There is no fee or formal government fiat or other official recognition needed to use the law's protections. Either way, with or without a patent, a business owner is remiss in failing to take advantage of the protections offered by the Trade Secrets Act.

Next, the most important definition—what this book all is about:

"Patent." A patent is an exclusive right, for a limited period of time, to the patent owner to deny others the right to manufacture, use, or offer for sale the patent owner's invention. The Office grants this right to the patent owner "in exchange for public disclosure" of the invention. In other words, once your invention is made public, the Office has given others the opportunity to steal your idea. There are a lot of guys who spend time reviewing new patent applications, then deciding to infringe, to steal your idea, or even to invalidate your patent. More about these bad guys later, but be aware that once you are given a patent, your preciously won grant, your brand new patent is no longer a secret. After eighteen months, your invention will be made public unless you filed a "Non-Publication Request" form. There is no fee for filing this form, and it cannot be denied. By all means file this form when you file your patent application. If you don't, your invention will be made public eighteen months after your filing, even though your patent may not have been issued. When considering the years, literally, it typically takes for inventions to be patented (more about this later) your secret will be kept. If you're granted a patent, your "Non-Publication" filing will be voided automatically, of course. But if your patent application is denied, your application will remain a secret forever.

I don't understand why, but the Office makes it easy for infringers and others who would do inventors harm, including you, by letting them learn about new inventions, but it does this by publishing every Tuesday in the Official Gazette* (eOG:P) a summary of newly patented inventions. At least, take some comfort in knowing that the Office won't tell anyone about your application, your filing; it's when you've actually been granted it that the cat's out of the bag.

---

* Take some time to browse through this magazine, noting official announcements about Office policies, press releases, FAQs for patents and trademarks, re-examinations, and most important, the complete file of inventions now in the "public domain, " i.e., those filed eighteen months ago, including those whose patents hadn't been granted, but who, surprisingly, neglected to file the "Non-Publication Request" form.

There are three types of patents:

1. **UTILITY** patents, the most common type of patent, for the invention or discovery of any new and useful process, machine, article of manufacture, composition of matter, or any useful improvement thereof. The key word is "Process," that is, an act or method involving industrial or technical activities, e.g., a new way of refining, smelting, curing, brewing, conversion, transforming, etc., or otherwise "usefully" improving the process.

2. **DESIGN** patents for a new, original and ornamental design for a manufactured article, e.g., jewelry, furniture, computer icon, etc. Design patents have nothing to do with usefulness, i.e., utility. Thus the unique shape of the Coca Cola bottle serves only to distinguish it from other beverage containers. To qualify for a Design patent, the design must be uniquely different from other similar designs. Thus the widely used Arial font earned a Design patent even though it differs from its virtual twin, the Helvetica font— developed years before—only by the "slant" of a few characters, e.g., t, r, and f. The Arial only "looks" a little different, but is sufficiently different to have earned its own patent.

3. **PLANT** patents for discovery or reproduction of any distinct and new variety of plant, e.g., a new rose variety, a different type of apple, a wheat that is rust-resistant, or any other plant that is significantly different in at least one characteristic. "Asexual reproduction" is the primary standard in the granting of a Plant patent. Further, as in all other patents, it must

be "non-obvious to someone skilled in the arts." Other than that, Plant patent applications are reviewed on a "case by case" basis, as the Office says.

Generally…

A patent cannot be obtained for "a law of nature, physical phenomena, abstract ideas, a new mineral or plant found in the wild, inventions used solely in the utilization of special nuclear material, printed matter, or human beings." Thus the Supreme Court recently ruled that human genes are ineligible for patenting.

Under current law, a patent is the grant or concession for twenty years for Utility and Plant patents, (for Design patents it's fourteen years) to an exclusive right to manufacture, use or sell an invention, provided maintenance fees are paid at prescribed intervals. The clock is set on the date on which the application was filed in the United States. You will lose your patent protection after the fourteenth year for a Design Patent, and after the twentieth year for Utility and Plant Patents. You will also lose it during the "protection period" if you fail to pay the prescribed maintenance fees, which are due before the 4th, 8th, and 12th years after the issue date. "Small Entities" (individuals, small businesses, non-profit organizations) currently pay $500 for a Filing Fee, and then are charged an additional $700 for an Issue Fee for an approved application. Other fees may be charged, including late fees during the review process. (See Appendix C for Fee Schedule, effective March 19, 2013).

Although it is not commonly known, you may also lose your patent if, under the Patent Reexamination Laws, "a substantial new question of patentability … of your patent … is presented." In other words, anybody (including the notorious patent trolls, of which more later) can file a request—paying the $17,750 fee and offering proof— for reexamination of your patent, not only once, but many times, on different grounds, challenging the validity of your patent. No matter the hollow assurances of the law that there must be "maximum respect for the reexamined patent … to minimize the costs and complexities of the reexamination process," you'll still have to pay your attorney his fee—don't enter this swamp without one— which could be substantial, depending on the issues, almost as if you were prosecuting your patent for the first time.

Following are the major hurdles you must overcome in order to receive your patent. Use it as a check list to review before you start spending money and time pursuing your dream. Then review it again.

Is It Useful?

Will it serve to better a condition, to improve it, to make it easier for an operation to be performed? "Useful" includes operativeness. Does it work? Does it actually do the job? If these conditions cannot be proved or shown, the patent will not be granted. Pluses would include: it does so easily, economically; it's earth-friendly; it betters conditions for mankind.

Is It Novel or New?

If your idea has been printed or otherwise broadcast in any publication throughout the world, your application will be denied. In other words, an invention cannot be patented if: "(a) the invention was known or used by others in the United States, or patented and described in a printed publication in the United States or a foreign country before…" your patent application was filed. Or, (b) "the invention was patented or described in a printed publication in this or a foreign country or on sale in the United States," then it will be denied—unless you can show that it's different in some important way. For example, that your product is activated by a chemical that's different than the one the other products use, that light rather than electricity is the activating force, that it uses a different material that is demonstrably safer or more economical to use than the other products, that it's more "earth-friendly." All of the foregoing examples, among others, would be grounds for granting a patent because that would make it "the first of its kind."

Is It Non-Obvious?

It must be sufficiently different from the most nearly similar thing—such that it will seem to be non-obvious to someone with "ordinary skill in the area of technology" involved. In other words, an "obvious" idea won't qualify, won't result in a patent. Admittedly, this principle is difficult to understand and apply in practice. Try to think of it this way: Aren't machines activated by an electric source? Hasn't that always been the recognized means of operating machines? Clearly, the answer is yes—then wouldn't activation by a chemical or by light be "non-obvious"? Or: Clothing had always been made of "natural" products like wool or other animal products until the synthetic fiber, rayon, was developed and patented. What about "natural" meat products, i.e., from animals, for human consumption? Chemically treated soybeans have now been accepted and patented as a non-obvious meat substitute. Vacuum tubes played a key part in the development of the radio and TV broadcasting industry until the non-obvious transistor and other solid state devices supplanted them as more reliable and cheaper electronic devices.

Okay, you've reviewed the foregoing criteria. You're sure you've met them. Now, what are your chances for getting a Utility Patent, the most common type, for your product? Feeling lucky? Fewer than fifty **Feeling** percent who file for Utility patents actually receive a patent. In **Lucky?** 2011, 503,582 total applications were filed; 224,505 (45%) were granted. In 2010, 490, 226 filed; 219,614 (45%) were granted. The same ratio prevailed for prior years. For other two patent types—Design and Plant—the ratio is much better, more than 70%. (See Appendix B for the latest Office report—from 1963-2011, on this subject.)

As you prepare to file, make sure that you have a complete record of everything you did leading up to the development of your invention. Use a log book, a bound book—not loose-leafed—of every event dealing with your invention. Have pages notarized from time to time, especially events that show "due diligence," in following through to the final product. Always include dates, e.g., when did you get the idea? Why (and when) did you think this was important? Were there other important considerations that occurred to you at the time? What other products are there like yours? How and when did you find out about them? What problems do these products, including yours, claim to solve? What uses do they serve? Are there seminal, important, documents bearing on the technology involved? (Give citations). Why is yours different than the others? What trials and errors did you experience in developing your product? Over what period of time? Which trial was successful? How did you know it would be successful under various conditions? Did someone help you? What was the nature of his assistance? Did he sign a non-disclosure agreement? **HE SHOULD HAVE.** Include the agreement in your log book

When your patent is issued, what are your plans?

What will you do with it? In other words, satisfy yourself that you have a complete record of all the events significantly bearing on the outcome, including plans for the future. This is because you'll need to satisfy the patent examiner and another important guy, your attorney, if you've hired one, who'll need all the help you can give him in "prosecuting" (that's the operative word because that's what is actually involved in your duels with the patent examiner) your case to the Office. If you're "prosecuting" alone, without an attorney, you'll still need to be armed with all the facts you can muster in answering the patent examiner's questions, and you can be sure there'll be many. In the event the Office receives two patent applications for the same invention, the cases go into

"Derivation Proceedings" to determine whether the inventor who filed his application first obtained information from the inventor who filed later, without his approval. This is where your meticulously prepared log book will prove your case.

Here's something you need to know about the recent law, "America Invents," which takes new major steps in protecting inventors. Under the new "First to File" (It had been "First to Invent.") procedures, an inventor has a grace period of a year of protection dating from his filing. If another inventor described a product identical to yours, or exhibited it at a trade show during the year or less prior to your filing, his product "shall not be prior art," under the new law. In other words, your patent cannot be rejected because it was the same as the other product, provided the "other product" had been commonly known a year or less prior to your filing.

CHAPTER TWO

BUYER BEWARE

"The Constitution never sanctioned the patenting of gadgets.
Patents serve a higher end—the advance of science."
– Supreme Court Justice, William Douglas

By now you've convinced yourself that you've got a great idea and are prepared to go forward with it, to spend money and time getting your patented product to market. Is there an economical, easy, fast way of doing this, being guided by experts? You've been looking at those attractive siren songs on the Internet and TV telling you how to turn your idea into money: "Let us show you how easy it is to make money with your invention..." "Take the first step with our help on the road to riches ..." "Call us for our FREE consultation and patent services..." etc.

Before contacting any of these organizations, read the following testimony given, in part, by former Deputy Commissioner Michael Kirk, to a U.S. Senate Subcommittee. Although Kirk spoke in September, 1994, much of it is still relevant in describing the abuses of Invention Development Organizations, as they are called, who still prey upon on unsuspecting inventors.

"Mr. Chairman and Members of the Subcommittee:
For the United States to compete in the world economy, we must

create new products and new processes, indeed, entirely new markets. The reason is clear: increasingly, the value of goods and services is made up of creative new products and methods of manufacture. We as a nation can compete and win in this environment if we properly nurture our creative community. One of the ways we can do this is by protecting the fruits of creative labor through the intellectual property laws of the United States. It is these laws that recognize and reward creativity to provide an incentive to create.

An important pillar in our intellectual property system is the United States Patent and Trademark Office ("USPTO"). The USPTO receives applications for patents—over 180,000 last year. If an application meets the requirements of the patent laws, protection is granted. Applicants rely upon the USPTO to operate in an honest, open and speedy fashion. Because of the complicated nature of the process, applicants also typically rely upon the help of registered attorneys or agents to represent their interests before the USPTO.

A number of inventors, however, confronted with a process they do not understand for protecting and commercializing their inventions, turn to invention development organizations. An invention development organization may offer to do several or all of the following: evaluate an invention, assist in its development, protect it, and promote it in order that it may be licensed or sold. While many of these organizations are legitimate, the unscrupulous ones take advantage of untutored inventors, requiring large sums of money up-front from inventors while providing no real service in return. In the process, some misuse and misrepresent the services of the USPTO, falsely claiming to obtain appropriate protection for their clients from our Office. Moreover, in addition to wasting their clients' money, some of their activities actually harm their clients' interests.

The result of the activities of unscrupulous invention development organizations goes beyond a loss of money and even beyond a loss of protection for inventions. If left unchecked, it could result in a general loss of confidence by independent inventors in the very institution designed to encourage invention."

———

Continuing the Commissioner's testimony… "The key to a proper result at each step along the road from an invention to the marketplace is that the inventor be adequately informed about the protection available, how it is obtained, what the scope of protection is, and how the invention can be commercialized. Inventors having knowledge about the process are more likely to be prudent and less likely to be misled into paying for a service they do not need or into doing something against their interests.

THE GENERAL NATURE OF THE PROBLEM

In an article dated February 11, 1994, the Wall Street Journal estimated that invention development organizations take in about $100 million per year. But for the activities of some of their members, the size of the industry would not be of concern. Indeed, if they all operated in an honest, open, and effective manner, we would hope for an increase in the size of the industry.

Unscrupulous invention development organizations attract inventors with tales of huge royalties, take their money, and provide little, if any, real service. Indeed, they often do real harm to the interests of those they are ostensibly trying to help. I want to emphasize that there are reputable invention development organizations. The basic problem for the unwary inventor, however, is that he or she lacks sufficient knowledge about the process of protecting inventions and information about invention development organizations to be able to distinguish between the good and the bad.

After the inventor/customer discloses an invention, the organization will purport to conduct a patent search. The unscrupulous ones rarely do so with the rigor required, and invariably will report that the invention is valuable and patentable. Based upon this favorable report, the customer is lured into paying for invention evaluation, marketing, brokering, and/or promotion services and possibly for the preparation of a patent application."

———

Concluding the Commissioner's testimony… "We have also seen evidence of organizations that file design patent applications in cases in which a utility patent application should be filed. Recalling the example of the self-watering mechanism for a flower pot, an organization may file for design patent protection directed to a silhouette of the pot or a floral picture on the surface of the pot. While a design patent may prevent others from making a pot having the same silhouette or floral picture, it will provide no protection for the self-watering mechanism. Taking this one step further, we have seen instances of a design patent being applied for where the design, such as a floral picture, has been added by someone other than the inventor by a draftsman hired by the invention development organization, for example. This addition by someone other than the inventor throws into question the validity of any design patent that may issue from the application. This is because the named inventor (the customer of the invention development organization) did not invent the design claimed in the design patent.

Thus, the protection sought through a design patent is often wholly inappropriate in light of the nature of the invention. Moreover, even if a design patent is granted by the USPTO, the manner in which it was prosecuted before the USPTO may render it unenforceable in court.

We have seen instances where inventors have lost domestic and foreign patent rights due to the way these organizations operate. This may happen in a number of ways. For example, the original invention may be of the structure or function of a product—an invention susceptible to utility patent protection. Nonetheless, the organization may arrange for a design patent application to be filed that only discloses the surface of the product. Because it is a design patent application, it need not disclose any structural or functional details of the product.

Once the organization begins its activities to market the invention, its disclosure frequently bar utility patent protection. This is for two related reasons. First, U.S. law bars utility patent protection to those who offer to sell their invention more than one year prior to filing for patent protection. Second, the only patent application that has been filed—the design patent application—did not describe or claim the structural or functional details for which utility patent protection could be sought. Under most foreign systems this bar is even stricter—patent protection is barred for public disclosure of the invention at any time prior to the filing date.

As a result, customers of unscrupulous invention development companies are left with a patent that is not useful, which was not competently prepared, and for which a high price has been paid. Frequently, no patent application will be prepared and filed and the inventor will have no protection. Moreover, if an application is improperly prepared, the applicant can and will obtain less protection than he or she needs or may be entitled to. Indeed, a patent application that is poorly drafted or not timely filed may result in a loss of rights in the invention, not only in the United States but also abroad.

Whether or not these organizations provide real value to their customers cannot be accurately measured. However, in providing information required by some states, some invention development organizations have admitted they had few customers that earned more from an invention than was paid to the organization, sometimes as few as 1 out of 10,000.

When patent applications are filed in the USPTO on behalf of customers of invention development organizations, they are often submitted by practitioners admitted to practice before the USPTO. By statute, the USPTO registers these practitioners to practice before it. The USPTO regularly receives complaints from inventors about these practitioners, including lack of information about the prosecution and status of their applications. When complaints by customers are filed, or irregularities are identified by the USPTO, we investigate registered practitioners, some of whom have been affiliated with invention development organizations. The costs of these investigations are borne by all patent applicants whose fees fund the USPTO.

There are intangible costs as well, the most troublesome being a loss of confidence by independent inventors in a patent system they see as frustrating, rather than promoting, the recognition and protection of their ideas.

––––––

We strongly believe that the problem of unscrupulous invention development organizations is persistent and growing. It is a problem that threatens the independent inventors that are directly affected by the unscrupulous invention development organizations. Further, it tarnishes the reputation of the many legitimate organizations dedicated to assisting inventors and undermines the integrity of the system for the protection and therefore encouragement of inventions. Finally, it tarnishes the reputation of the USPTO."

––––––

Commissioner Kirk urges inventors to be aware that "only a registered attorney is competent to advise them about their legal rights in patents, trademarks, and copyrights... In addition, inventors should ask the invention development organization to disclose: " the 'success rate' – the number of clients a company has had that have made more money from their invention than they paid for services rendered; and the "turn-down rate" – the number of prospective customers a company has told they will not represent, either because the invention was not protectable or that it was not, in their opinion, commercially viable in comparison to the total number that have come through the door.

Presumably, if the success and turn-down rates of a given invention development company are low, no rational inventor will want to use its services. It appears that the true success rate and turn-down rate among the unscrupulous segment of the industry today is virtually zero.

Finally, Kirk says that a 'large sum of money up front for future services" is a red flag.

———

In his testimony, Commissioner Kirk failed to explain why the invention development organization filed a Design patent, rather than a Utility patent, which would have been more appropriate for the product in question. It was simpler and easier to apply for a Design patent, compared with a Utility patent. More important: 70% of Design patent applications are approved compared with less than the 50% approval rate for Utility applicants—hence the unscrupulous organization here was able to offer specious proof that it had successfully met its obligations to the client, and probably was able to collect more money from the victim.

Before taking any chances with the organizations Deputy Commissioner Kirk warns us about, simply avail yourself of the considerable amount of free or low cost advice available on patenting and other forms of intellectual property protection. A partial listing, available through Google, includes "National Society of Inventors," with branches throughout the country. Their motto, "Inventors Helping Inventors." Inventors Digest is a monthly magazine "committed to educate and inspire independent and professional innovators." Others include the Inventors Network "which helps people develop and pursue their ideas by bringing them the latest and greatest information on how to bring a product or idea to the marketplace." Still other inventor organizations include the United Inventor Association of the USA. These and others are eager to share their knowledge about the invention/patent process. Consider joining an inventor group in your area, attending their meetings. If you decide that you still want the help of an invention development organization, you'll receive impartial advice from members of some of the foregoing inventor groups.

The federal government has a program to help inventors whose ideas or products have been stolen or otherwise compromised. The government has responded to the need for protecting American inventors from the recent proliferation overseas of the counterfeiting of American products. For information about how inventors may be protected, what types of protection are available, and how to file, go to www.stopfakes.gov.

CHAPTER THREE

THE SEARCH

"If you build a better mousetrap
you will catch better mice."
– George Gobel

Your invention is patentable, you think. Should you quit your day job? Are you going to go ahead with your patent application? Before you spend even that relatively minor amount, compared to the much greater amount you'll spend later on if you decide to go "all the way," you need to know whether somebody else has thought of your idea. You'll want to guard against surprises—I mean the unpleasant ones—the ones that always have a way of showing up after you've spent a lot of money and time.

Since its founding under the first Federal Patent Act in 1790, more than eight million patents have been issued by the Office. Add to that the patents issued by other countries for their citizens, and finally, add inventions or concepts described in publicly available documents both in the United States and abroad. Is yours the really "something new under the sun?" Probably not. Chances are yours is not really that new—it's more likely to be a new, different or unique application of an existing technology. It's unlikely to be the "next big thing" when considering the patents granted in the United States and in all other countries with patent laws, not including ideas and concepts described in publicly available documents since the dawn of the written word, pre-dating the Gutenberg Bible.

There are firms that offer to do a patent search, and more, for a "low-ball" flat fee—usually $100 to $200—because "we know how to do it." Stay clear of these guys, no matter what promises they make. Theirs is simply a bait and switch game.

Then, there's doing it yourself. I recommend this course of action, at least to begin with, if for no other reason than to gain some understanding of the wild world of patents. Besides, taking this course of action may lead to some changes in your product you may decide to make that will distinguish it from other products in your subject matter area, thereby facilitating the patent process. Equally important, as you survey the field, there may be some changes you'll want to make, for example, that will lessen manufacturing costs, speed up/simplify the assembly, improve its appearance, make it more "earth-friendly," or make it easier to use. In addition, doing it yourself will give you a good overview of the technology and terminology of the domain you've now entered. Finally, you may also turn up with an unpleasant surprise—someone has already patented a product exactly like yours, and you won't be able to make any changes to your product to distinguish it from the other one without doing violence to your fundamental concept. This is the fork in the road that Yogi Berra told us about.

Google, with its 8,000,000 patents, offers free help in your search. The Office also offers patent search assistance in its "Step By Step " tutorial. For more Office assistance contact their Scientific and Technical Information Center. Their phone is 800-786-9199. The Center currently receives 150 calls per day, so expect to wait. Another resource is the library in your state that has been set up as a Patent and Trademark Depository (PTRC). The Office has set up at least one such facility in every state. Google "Patent & Trademark Depositories" to find the library(ies) in your state. Even though individuals at these facilities are given frequent training by the headquarters staff, you'll get more help by going to Alexandria—or to one of the "Satellite" offices, as explained below—with its 120,000 volumes of scientific and technical books in various languages, and 90,000 bound volumes of periodicals devoted to science and technology, plus many other sources of information bearing on virtually all of human knowledge in science and technology.

"Satellite" branch Offices have recently been set up in Detroit, Pittsburgh, Dallas/Ft. Worth, and in San Jose , California. These branches are staffed with patent examiners just like those in the Office's headquarters in Alexandria, Virginia. In other words you'll be able to talk directly with a patent examiner, or at least another subject matter specialist, to help in your search. These four satellites were set up under the 2011 "America Invents" law, but other satellites may be set up in the future, if for no other reason than to help with the unemployment problem in more cities. But a search at your State library may be all you need to make a preliminary search.

The Office offers a broad range of assistance about the patent process, including help in filling out forms. To avoid being passed from one Office employee to another, it's best to ask for the "Inventor Assistance Program," (IAC) the most knowledgeable group. (800-786-9199.) They're available only from 8:30 AM – 5:00 PM ECT.

The following helpful pamphlets are available from the Office:

- Provisional Application for Patent
- Guide to Filing a Utility Patent Application
- Guide to Filing a Design Patent Application
- General Information about Plant Patents 35 U.S.C. 161

To order, call the IAC or key in www.uspto.gove/web/patents/guides/htm to order these pamphlets.

Okay, you're done your own patent search, so you now have a better handle on the patent game and understand something about its complexities. But don't assume, however, that your search has uncovered all possible "prior arts" products like yours. The question: should you hire a patent attorney to do your search? Don't be misled by the assurances of some attorneys urging you to let them do it because they'll "put it all together in a package" deal that will take care of everything for you—the search, the filing, everything from A-Z. Why spend a higher fee for a lawyer to do your search when you can hire a "patent searcher," a Patent Agent, one who specializes in this activity, to do a good job for you for a lesser amount.

For a complete listing of Patent Agents ("searchers") go to Listing of Active Patent Attorneys and Agents by Geographic Region , attorneys and agents licensed by the Office. Your best bet would be with an Agent in the Washington area. All of the source materials needed by the searcher for a thoroughgoing search are literally close at hand. Also available are the patent examiners in your subject matter area whom your searcher can consult, adding to its thoroughness. These searchers typically are specialists who "really" know how to do it. Further, a good search will result in a good application, also with lower attorney's fee, if you've hired one—he has had to spend correspondingly less time preparing the application. A comprehensive search should cost less than $1,000. An attorney would probably charge more than a well qualified Agent for the same service.

A good search will disclose and list all prior arts, even those bearing remotely on your concept—that includes copies of documents about concepts in your subject matter area. Your application will be deemed fraudulent otherwise. (If the Office finds, conducting its own search, that you failed to cite a prior art similar or identical to your product, your application will be rejected.) Check references. If the Agent doesn't offer a confidentiality agreement, proceed carefully. Don't hire one without an iron-clad agreement.

Hopefully, the search has given you a green light, free of conflict or overlap with prior arts, to go ahead with the selection of your patent attorney—a must is one registered to practice with the Office. Refer to the above list of registered attorneys in making your selection. Plan on paying $200-$500 (or more) per hour, depending on whether you're dealing with a "big name" firm or other firms. My own experience had always been to sign up with one of the "other" firms who don't have to contend with their overheads, including window dressing, of the big shots. The table below will give you an idea of what to expect to pay an attorney:

TYPE OF INVENTION	EXAMPLES	ATTORNEYS FEES
Extremely Simple	Electric switch; coat hanger; paper clips; diapers; earmuffs;ice cube tray	$5,000 to $7000
Relatively Simple	Board game, umbrella, retractable dog leash, belt clip for cellphone, toothbrush, flashlight	$7,000 to $9,000
Minimally Complex	Power hand tool, lawn mower, camera, cell phone, microwave oven	$9,000 to $10,000
Moderately Complex	Riding lawn mower, simple RFID devices, basic solar concentrator	$10,000 to $12,000
Relatively Complex	Shock absorbing prosthetic device, basic to moderate software/systems, business methods	$12,000 to $15,000
Highly Complex	MRI Scanner, PCR, telecommunications networking systems, complex software/systems, satellite technologies	$15,000 +

Courtesy: IP Watchdog, Inc

My working assumption: I would pay $10,000 for a fully qualified, registered attorney to guide me through the rocks and shoals, from filing to approval. I was able to pay less when I was able to show a truly comprehensive search combined with a detailed written rundown of everything leading up to product development—my "log book." Good drawings helped too. An in-person meeting with your attorney will tell you whether there's good chemistry. Ask for patent applications he's filed for his clients, including their names. Follow up with them. If he's been in court, Lexis will give you information about the case, including his representation. I'm making the assumption here that you're going to go all the way and have decided to fully commit yourself. Otherwise, don't spend any more money and time. Above all, don't make the mistake of telling your attorney, "I've got this idea."

CHAPTER FOUR

THE APPLICATION

"Necessity is the mother of invention. The father is unknown."

Disclosure Document

Until recently, a filing could be made with a Disclosure Document. This program, recently discontinued by the Office, had been the source of misunderstandings and confusion. In no way had it offered any protection to the inventor. Patent filings are now made only with either a Provisional Application or a Non-Provisional Application. You'll know you're being set up if one of the Invention Development Organizations offers to file a Disclosure Document for you.

Patents these days are filed electronically—"Snail Mail" is no longer the primary means of sending patent applications. The Office has instituted an electronic filing program called EFS-Web. Hand filing a Provisional application by mail will cost you $395 for a "Non-Small Entity." Small entities (individuals, small businesses, non-profit organizations) are charged $195 for hand-filing. In addition to the $195 and $390 fees for a Provisional Utility patent, applicants must pay the fees for search and examination. (See Appendix C). Besides saving you money, by filing electronically you'll get an electronic receipt fixing the date and time of your application as the "first" patent owner. In the meantime, secure servers at the Office will safeguard your file. Contact the Office for help about EFS-Web at 866-217-9197.

The Provisional Application

You can beat someone else in the race for patent ownership by filing a Provisional Application for either a Utility patent or a Plant patent. Design patents applications may be made only by a Non-Provisional application. (See below.) You can do this with the assurance that your invention will not be stolen, just as if you had filed a Non-Provisional Application; and with the further assurance that it will never be made public. You can now lawfully use the term "Patent Pending" on all documents about your invention. Be sure to also mark your products "Patent Pending," simply to give notice to possible infringers that they're asking for trouble if they plan on producing your product, or in any way claiming ownership. After you've received your patent, the law requires you remove this notice.[*] The Office makes it easy for you to file provisionally, relieving you of the need for detailing your claims, disclosing prior arts, and everything else required in the Non-Provisional filing. Even though formal review of a Non-Provisional application doesn't take place, be sure that your Provisional Application is done with the same level of care as your subsequent Non-Provisional filing. See the Office "fill-in" form SB 16 for this type of application.

Among reasons for filing a Provisional application: You want to beat a possible competitor who has a similar product. Or you may want to shop around for a possible licensee for your product, then decide whether it's worth your while to go ahead with a Non-Provisional application. Unless there are other compelling reasons, why file twice, spending the

---

[*] If you haven't applied for a patent, don't assume that you can use "Patent Pending" on either your documents or on your products. Doing so will invite the severe penalties of "unlawful use" under the law. You can be fined $500 for each occurrence, meaning for each of your products so imprinted, made "with intent to deceive the public."

extra $195 and time, first for the Provisional, then again for the later filing. In any case, you must file the Non-Provisional in less than a year, otherwise you lose your place in line.

The Non-Provisional Application

More than half a million Patent applications of all types are received annually by the Office. Who said America lacks for innovative, imaginative people? That seems to be the essence of the American spirit, that there's always a less costly way, a better way, and lately, a "greener way," one that will reduce drudgery, one that will save lives or otherwise improve the human condition.

At present, 11,531 employees, including 7,935 patent examiners, work for the Office. Remember to treat the examiner with the courtesy and respect he deserves. Information about your application will be safe in his hands, as employees of the Office are expressly prohibited from applying for patents themselves, except that examiners work one day a week at home, under established Office policy. This raises important questions about the security of the applications being reviewed in a "home" environment—also about the adequacy of supervision, including the quality of interviews with "stakeholders," as the Office calls patent applicants. In person interviews with patent applicants is difficult in these circumstances. Try to be patient as you await the first response from the Office. The Office claims that the typical wait time for the first Office action is two to three months, contrary to the experience claimed by many applicants. "Much longer" is a common complaint. More important is the time involved before the patent is finally granted. Former Chief Judge Paul B. Michel of the U.S. Court of Appeals for the Federal Circuit, which receives appeals on patent cases, has said, "Inventors must wait three years, on average, and often much longer…Some areas of technology may take as much as six years," meaning that much time will be shaved from the patent's period of protection because the patent clock starts ticking on the date the application was filed.

The Judge commented about the undue time delays in a speech he gave to the Office patent examiners in April, 2010. Since then a number of changes have taken place in the Office under "America Invents," major legislation passed in 2011—no doubt, in response to Judge Michel's criticisms, as well as complaints by many others regarding the "dysfunctional" Patent Office, the Judge's term.

In April, 2010, when the Judge spoke, there were 700,000 patent applications waiting for "first Office Action." There has been some reduction—600,000 plus before "first Office action." The "dependency" (waiting period before the application is approved) is still three years. A major effort is currently underway in the Office to reduce both the backlog and the dependency period to two years.

Another backlog has yet to be effectively addressed: the Office admits, "that as the number of patent examiners has grown, the number of new ex parte appeals (by "one judge," in this case, the Patent Trial And Appeal Board) there has been a corresponding growth in the time it takes for an appellant to receive a decision on an ex parte appeal; it has doubled " in the past two years—from 2010-2012. In other words, a patent applicant may have to wait at least five years in these circumstances before he owns his patent, free and clear (unless he's challenged by patent trolls or other bad guys, of which more later).

Our Dysfunctional Patent Office

Another recent change: the Office has set up the "Track One" program, whereby the first 10,000 applicants in each fiscal year will get "prioritized" treatment upon payment of $2,400 for a "small entity," and $4,800 for a "non-small entity." Track One is open to Utility and Plant applications, but not to Design applications. The application may not exceed four independent claims and thirty total claims. Applicants with a currently pending Non-Provisional application are eligible to submit a new continuing application for inclusion in Track One . The Office claims that "final disposition will be made, on average, within twelve months of the date that prioritized status was granted."

The Office "strongly recommends" that applicants use the Office certification and request form PTO/SB/424 to request prioritized examination…The form is available on EFS-W and on the Office's Internet Web site. Call 866-217-9197 for assistance in downloading this form. "Failure to use form PTO/SB/424 could result in the Office not recognizing the request or delays in processing the request."

All other applicants are in "Track Two," in other words, with a dependency period of three plus years, not including ex parte appeals. Add another two plus years if you appeal the examiner's adverse decision.

You may be able to speed up the process by a few months or more by claiming and proving age (65 plus) or poor health. That means your application will be placed ahead of younger or healthier applicants.

The application, whether Provisional or Non-Provisional, must observe the following order, exactly (cf. Cooper Patent, Appendix D):

"Application transmittal form

(a)		Fee transmittal form
(b)		Application data sheet
(c)		Specifications
	(1)	Title of the invention
	(2)	Cross reference to related applications (if any). Related applications may be listed on an application data sheet, either instead of or together with being listed in the specification.
	(3)	Statement of federally sponsored research/development (if any).

(4)   The names of the parties to a joint research agreement if the claimed invention was made as a result of activities within the scope of a joint research agreement.

(5)   Reference to a "Sequence Listing," a table or a computer program listing appendix submitted on a compact disc and an incorporation by reference of the material on the compact disc. The total number of compact discs, including duplicates, and the files on each compact disc shall be specified.

(6)   Background of the invention.

(7)   Brief summary of the invention.

(8)   Brief description of the invention.

(9)   Brief description of the several views of the drawing (if any).

(10)   A claim or claims.

(11)   Abstract of the disclosure.

(12)   Sequence listing (if any).

(d)   Drawings.

Executed oath or declaration."

What Can Go Wrong—And It Will

Okay, you've filed your application. Smooth sailing now? No way—this is where the fun starts. Following is a partial list of the reasons your application will bounce back to you.

- **FORMAT**. You haven't followed the standard form, e.g., missing Abstract, Drawing, Background. The Office is also persnickety about the right size of paper—and only one side may be used.
- Where's the **OATH**?

- Where's the **FILING FEE**?
- **MINIMUM REQUIREMENTS**. You didn't include a specification and at least one claim.
- **INDEFINITENESS**. The examiner doesn't understand your claim. He means you don't know how to write good English, or at least the jargon used by the Office.
- **TOO MANY INVENTIONS**. Sorry, only one invention per application.
- **TOO BROAD**. Your patent covers too much. You need to overcome his "Request for Restrictions" in order to keep your patent broad. (This is where the application process begins to drag. You want your application to cover the world; the patent examiner wants to confine it narrowly, to circumscribe it).

  "The Patent Office is the mother-in-law of invention."

- **IT LACKS NOVELTY**. You need to show why your invention is different from the prior art.
- **IT LACKS NON-OBVIOUSNESS**. You need to prove—to paraphrase the Patent Law—that your invention wouldn't have been "obvious … to a person having ordinary skill…in the art…"
- **YOUR DRAWING DOESN'T SUPPORT YOUR CLAIMS**. You shouldn't have had your know-it-all brother-in-law do the drawing. Hire a professional artist.
- **DESCRIPTION DOESN'T SHOW THE MANNER AND PROCESS OF USING IT**. Try to find published articles about products like yours showing that it is in common use and is easily used by the people working in the same technology. Can you find someone who works in the same or related field who'll furnish a written statement supporting your claim?

The questions, including objections (termed "OA," Office Actions), like the foregoing, raised by the patent examiner, must be answered promptly, otherwise risking "Abandonment," then having to pay a fee to activate it.

Profit by my mistakes. At first, I tried to do it all by myself—search and Non-Provisional filing, everything, with the result that I wasted a lot of time fencing with the patent examiners, leaving me with the feeling (shared by all applicants) that the Office staff is your adversary, doing everything they can to prevent you from getting your patent. I finally decided to hire a well-recommended, registered patent attorney, keeping my place in line with my original filing. All he had to do was to file some amendments. Bear in mind that you will now be cut off from any direct connection with the Office, as the Office will refuse to be in contact with both you and your attorney. Any comments you may have must be forwarded to the Office through your attorney.

You must disclose information about all products generally similar to yours ("prior arts")—or have your application rejected as fraudulent— and then explain why yours is different in a significant way. Following, taken from my patent for Smoke Detector Tester (See Appendix D) is representative of the questions dealing with prior arts. My patent search disclosed several products similar to mine—each using an aerosol product—but different from mine in a significant way. I needed to list them, otherwise having my application rejected as "fraudulent."

PATENT 349973. This product used the spray from an aerosol can to reduce the level of ozone (smog) in an enclosed space, thereby making the environment tolerable for plants and humans. While both products use aerosol sprays, the objectives are radically different—one to test smoke detectors (mine) and the other (theirs) presumably makes the air in the enclosed space "tolerable."

PATENT 3693401. While both products, mine and this one, used aerosol sprays to test smoke detectors, the applications were very different in the final application. To test smoke detectors with the other product, you needed first to place a metal housing over the smoke detector; then the aerosol was sprayed at the detector through the vents in the housing, presumably causing the alarm. The difference: their product added to the expense (for the housing) and time (for assembling it). On the other hand, my product dispensed with both the cost of the housing and the added time needed to test. Incidentally, after I obtained my patent, I contacted the owner of patent 3693401. He decided to use my product instead of his, giving me purchase orders over a number of years.

PATENT 3985868. This product was designed to give asthma sufferers relief by inhaling a powder discharged from an aerosol can. This matter was easily disposed of: the two products had different objectives.

Explaining why my product differed significantly from the other aerosol applications was relatively easy. But I had tough sledding, in particular, with questions of "Indefiniteness," "Non-Obviousness," "Too Broad," that the patent examiner had raised—questions commonly raised in patent applications, typically delaying the application process. The differences between my initial and final application were many and significant—all having to overcome the patent examiner's questions dealing with "Indefiniteness " and "Non-Obviousness," but especially the examiner's argument that my claims were "Too Broad. " (This is where that clicking sound you hear—your attorney's fee meter—begins to sound louder.)

Prepare yourself for the shock if the examiner tells you that your application has been rejected, and that his decision is "final." That's what he's trained to say. Back to the drawing board. You need to amend your claims, addressing each of his questions, then try again. Are you still hung up? Here's why an experienced attorney with a good track record is so important. He may know the examiner or his supervisor, or somebody higher up in the chain. Even if he doesn't have an "in" with anyone, he knows the pathways in the Patent Jungle. The matter may be cleared up in a telephone call or, if you're willing to pay his expenses, a trip to Washington may be the best way of smoothing the path to the final approval. Of course, you may be in luck if you're in or near one of the four "satellite" offices—Detroit, Dallas/Ft. Worth, Pittsburgh, San Jose, California. In all of this, it's a matter of your hopefully good judgment and the trust you've placed in your attorney when he assures you of ultimate success.

It's important to know about the **Manual of Patent Examining Procedure (MPEP)**. Click on www.uspto.gov.mpep— to become acquainted with the Bible of all matters concerning patents as administered by the Office. Don't try to be an expert on this arcane subject—that's why you've hired an attorney who is, or absolutely, must be. But at least be conversant with the terminology, the patent examiner's point of view, and the reasons for his questions. You'll also be better able to check on your attorney's progress or lack thereof in connection with the inevitable hangups that occur as you and he prosecute your application through the Office. Most important, with the Manual in hand, you and your attorney, as a team, can work together in coming up with good answers to the examiner's questions.

All else has failed, so you file an "Amendment Under 37 CFR 1.116 After Final Rejection," along with arguments supporting your claims. That means the Office will respond with an Advisory Action within three months. You can keep going in the face of more rejections, depending on your supply of patience, time, and money— for your attorney— including appealing directly to the Commissioner of Patents, then in the Office's Board of Patent Appeals and Interferences, finally on to the federal courts, including the Supreme Court. Do you really want to go through all this?

So you've finally decided to call an end to spending your money and time. Is that the end of the game? Wait. Was there a reason you had filed your application in the first place? Had you filed it in order to prevent someone else from stealing your idea? If so, you may still be successful in doing so if the other party intends to file an application for a similar product. You can block or at least delay his application by publishing your invention in a widely disseminated magazine or similar publication. The better course of action is to forever give up obtaining a patent for your invention and to request a "Statutory Invention Registration" (SIR—Form PTO/SB94) from the Office. Filing the SIR is the same as publishing it, making it a "prior art," completely blocking someone else from obtaining a patent on your product. What you've done by this filing is to declare your product to be in the public domain, free for anyone to make and sell. Of course, you've not prevented your competitor from manufacturing and selling his product. You've only denied him the benefit of patent protection.

But, surprise, surprise—you've been "Allowed." You're not surprised—you've known all along that you'd be Allowed, that's Office-speak, advising you that your patent has finally been approved. Along with the Notice of Allowance you've received a Fee Transmittal Form telling you that you need to pay the Office $300 for publication (See Appendix C), letting the world know about your invention.

You and your attorney should take a close look at your now-approved patent, this time with "new eyes." Although you are allowed some time (up to two years if you decide to broaden your claim), there's no point in delaying the correction of obvious typos and other glitches in the text. Making these changes with the Office's approval are now part of the file of your patent.

Decision, decisions ...

With your shiny new patent in hand, here's another set of decisions you'll need to make. Have you decided to manufacture and market the product yourself? Did you hire a market survey organization to help you define the market, especially its scope and type of customers? If the survey gave you "positive" answers, you'll need a business plan showing some hard numbers, including capital needed going in, direct and indirect costs, and, as always, making provision for "sudden costs," those costs that are unknown and unexpected, which always—I mean always—happen.

Now, do you want to bet the farm? Refinance your mortgage? Are you able and willing to work sixty to eighty hours or more per week, including weekends? Will the relationship with your wife and family be threatened by this demanding demon, your invention, which insists on having your undivided attention day and night? Who's going to take out the garbage? Are you prepared, for the rest of their lives, to face the bitter anger of relatives and friends who've paid for your failure? Remember: most business start-ups fail from lack of money.

Don't charge ahead without considering all the things that could go wrong (and you can be sure that something "wrong" will happen). Hang on to your day job until you're ready to go ahead, fully committed.

What about selling or leasing your product?

The Office lists your new patent for the $300.00 you had to pay any way. This will get the attention of a lot of people, including the bad guys I discuss later—also companies with a legitimate interest in your product. (Don't hold your breath). Google will give you a listing of companies looking for all kinds of products. Or, you can hire a broker who will find a company tailored to your needs—for sale or licensing, typically charging an upfront fee plus a royalty based on sales made by your buyer/licensee.

Before taking any action to sell or license, however, why not test the waters to find out whether you've found the goose and its progeny of golden eggs? Besides, if you do finally decide to sell or license, you'll be able to offer your prospective buyer a product with a credible track record, compared to a product without one—with the obvious price differential.

Let's assume that you've decided to license rather than sell your product outright. You see greater sales volume and greater exposure than you would be able to mount by yourself if you find a firm with an experienced, aggressive marketing team. Don't assume that there are scores of such firms waiting to sign up for your wonderful new product. Hopefully, you've found one whose interests mesh with yours. Your new licensee has now presented you with the agreement he has drawn up— remember, he's a pro in these matters. You can be sure the document will be slanted to favor his interests, not yours. You've spent a lot of money, including aggravation, getting your patent. After all, it's a license you're offering, not an outright sale. Also, the agreement will no doubt include a statement denying you the right to sue the licensee for patent infringement.

That's giving up a lot, so make sure you're protected against poor quality control, design changes, rotten treatment of customers, exaggerated claims about your product, among other things, that will do lasting harm to your chances of doing well in the marketplace if you decide to take your product back. What assurances do you have that you'll receive the royalties you're entitled to? In the last analysis, it's simply a matter of "chemistry;" your instincts that tell you whether or not the other party will deal fairly with you.

The ball is in your court. It's yours to decide whether to make or sell.

Okay, you've decided to "make," to undertake everything. You think you've seen that goose—you remember? You hope it doesn't turn out to be that obnoxious AFLAC duck. Only, don't make the mistake that some new patent owners make by going out to rent plush office space and fancy furniture, hiring that hotshot, high paid, salesman, and, of course, your own secretary (doesn't every executive have one?). Here's how to start small:

- Manufacturing. Your biggest initial outlay, even if you decide to farm it out. Expect to pay up-front.
- Advertising. What about getting free press coverage about an "exciting new product"? "Local boy makes good." Learn how to write good press releases.
- Marketing. Get over your fears. Start making cold calls.
- Office space. Remember that spare bedroom? Haven't you always wanted to tell your mother-in-law, "Sorry, it's taken." The kitchen table needs to be used for meals only a few hours a day—what about using the kitchen counter for meals?
- Staff. Your wife wants you to be successful, doesn't she? So she'll find time to become your staff. Your kids, too. No, not that know-it-all brother-in-law who'd insist upon a piece of the action, anyway.

Surprise! You've stumbled your way into what looks like it might be a successful venture after all. Your money, your hard work—and yes, your good luck, too—you've entered the market with the right product at the right time. So you've decided to do it all yourself, after all. Good luck! But watch your back, as I discuss in the next Chapter. You needn't go "bare," i.e., insuring yourself against possible infringers. There are insurance companies who'll reimburse

"Anything that won't sell, I won't invent."
– Thomas Edison

you for litigation expenses involved in enforcing your patent rights. Following is a table of sample premiums by industry—pretty expensive. Watch out for loopholes, too. How much is your patent worth?

INDUSTRY	POLICY LIMITS	SIR*	PREMIUM
Renewable Bioenergy Treatment Systems	$3M/$3M	$60K	$14K
Software Applications	$1M/$1M	$20K	$10K
Waste Water Treatment Industry	$1M/$1M	$20K	$15K
Data Management Services for Assets	$1M/$1M	$150K	$35K
Manufacturer of Installation Systems	$1M/$1M	$100K	$72K
Visual Search Technology	$2M/$2M	$40K	$29K
Ophthamolic Lens Technology	$2M/$2M	$40K	$62K
Environmental Products & Services	$1M/$1M	$50K	$$32K
Safety Equipment Wholesalers	$1M/$1M	$25K	$22K
Partition, Shelving & Locker Manufacturer	$2M/$2M	$40K	$29K
Medical Device Products	$2M/$2M	$150K	$31K
Industrial Equipment Manufacturer	$2M/$2M	$40K	$46K

*SIR (self-insured)

Per. . . Intellectual Property Insurance Services Corporation

If you win, all of your legal expenses are covered and you can keep as yours whatever damages or monetary awards the court decides. If you lose, your policy acts as a regular insurance policy—paying for legal expenses up to the maximum coverage.

CHAPTER FIVE

BAD GUYS

"I have been so constantly under the necessity of watching
the movements of the most unprincipled set of pirates
I have ever known, that all my time has been occupied in defense,
in putting evidence into something like legal shape that
I am the inventor of the Electro-Magnetic Telegraph."
(from a letter to his brother describing the challenge
of defending his patents (19 April, 1848)
– Samuel F. B. Morse

You're doing well. Purchase orders begin to roll in, and more after that when word gets around from customers telling others about your great new product, and now you have some money to spend for advertising. Your wife wants her kitchen table back, so you've begun looking for nearby, inexpensive office space—month to month. The local college registrar tells you, "Of course, we have students who want to work part-time."

Then, one day the mail brings a letter from the Office advising that someone has filed a "Request for Reexamination" of one of your claims. You're righteously angry—after all you've gone through, and now this! Why only one claim, you ask your lawyer, who tells you that's all that's necessary, provided the claimant paid the $17,750 fee and has satisfied the Office with the necessary documentation. Unfortunately, it's the one claim, he tells you, that significantly broadened your product's coverage, then asking, "Do you want to fight this?" "Yes," you sigh. He'd get back to you with his recommendations after having reviewed the claimant's documentation, but asked if you'd let him try to blow the claimant off by threatening him with a lawsuit.

"Didn't work," your lawyer tells you later, showing you a copy of the letter he received from the claimant, offering to withdraw the Reexamination request if you pay them an obscene amount for a license to their product; otherwise, they'll sue you for patent infringement.

"Calm down," your lawyer tells you. He's written to the Office citing The Doctrine of Equivalents, (See Appendix A) stating that the claimant's product differed from your product in only minor ways while producing the same results. There was infringement, your attorney pointed out; therefore, the claimant's request should be disallowed. "No," said the Office in reply to your lawyer. You must go through the same hoops as when you first applied for your patent.

What to do? Your lawyer tells you he's pretty sure the request for Re-examination is from a patent troll. "Patent troll? What's that?" you ask. Google or Wikipedia will tell you more about these bastards, also called patent extortionists, typically organizations having no intention of manufacturing the product themselves, often buying still active patents for a nominal price from bankrupt corporations or others, then using it as a weapon to threaten the original patent owners in a court action (now hapless defendants) in a bold attempt to extort money. And they've been phenomenally successful at their game. Wikipedia cited an analysis in 2011 reporting that "patent trolls cost U.S. individuals and companies direct costs totally 29 billion dollars in the United States alone." There are many instances, according to Wikipedia, in which the mere threat of litigation had led to settlement, including a royalty, by the beleaguered original patent owner. The U.S. Court Report for 2008, the latest year available for this report, states that, "For the 2,875 patent cases that were terminated in 2008, only 3.8% of the cases ever reached trial. 1,527 cases (52%) were terminated before pretrial, while 402 (14%) were terminated during or after trial, and 847 patent cases were terminated with no court action."

To Sue or Not

Following are the results of a survey made in 2011 by the American Intellectual Property Law Association of litigation expenses involving infringement of intellectual property, including patents:

AVERAGE (MEAN) INTELLECTUAL
PROPERTY LITIGATION COSTS*

Litigation Type	Amount In Controversy	End Of Discovery	Through Trial
Patent	<1M $1M-$25M >$25M	$490K $1.6M $3.6M	$916K $2.8M $6M
Trademark	<1M $1M-$25M > $25M	$214K $607K $1.2M	$401K $1M $2.2M
Copyright	<1M $1M-$25M >25M	$216K $543K $1.22M	$348K $932K $2M
Trade Secret Misappropriation	<1M $1M-$25M >25M	$303K $877K $1.9M	$521K $1.6M $3.2M

* The above costs exclude judgments and damages awarded.
With permission of American Intellectual Property Law Association. Data from AIPLA, 2011 Economic Services, pp 1-153-1-168, Arlington, VA

Add to these costs the time you and your staff must spend in depositions and in related matters, especially if the case winds up in court. This is why even large companies try to avoid litigating all the way through trial. In fact, the great majority of infringement cases are settled out of court.

Besides, who knows what a jury will do in cases involving the technical issues of a patent?

But at least one company, at this writing, has decided that it's not going to take it any more. The Wall Street Journal reported, on November 12, 2012, that "Cisco Goes On Offense Against Patent Trolls, Calls Them Criminal Shakedown Artists." Continuing, "Cisco has decided to go on offense against firms that purchase patents in bulk and sue other companies for infringement despite not making any products of their own by flat-out accusing them of breaking the law...the network-equipment maker has captured the attention of patent experts and lawyers across the country by filing strongly worded legal claims against two companies that buy up patents and seek to make money through licensing and litigation."

As the Cisco case is still in litigation, it isn't possible to know what effect its outcome will have on actions that may be taken by other companies by patent trolls, or indeed what action Congress or the Patent Office may decide to take.

So once again, it's your decision. The rewards could be great if you win—this time, at least, but maybe not when the next Reexamination comes up.

Whatthehell, you went through all those tribulations with the Office—also the money and time you spent. Aren't you entitled to do all you can to protect your rights?

In my case, I decided to sue a company, using a similar product (a copycat) that had repeatedly disparaged my product, Smoke Detector Tester. Their tactics were beginning to seriously affect my sales. After my repeated warnings to them, I took action. The Tester was one of the key assets of my corporation. I had no alternative but to pay my attorney the deposition costs and related expenses—increasingly heavy expenses as the proceedings dragged on. Finally. the defendant agreed to settle out of court—enough to recoup my legal fees plus "consequential damages."

I hope this booklet has been helpful in your quest for a patent. I've tried to guide you through the rocks and shoals, a rough trip, as you no doubt have discovered. As if the trip itself weren't difficult enough, you've had to contend with the sharks in the water, especially those who began circling you after you won your prize, a patent. At least, take some comfort in knowing that, having earned your patent, you've beaten the odds, two to one. Surely, that alone should beef up your confidence in contending with the bastards who aim to do you harm.

You've been fortunate in other respects, having been given a patent despite the vagaries and twists and turns of a "dysfunctional" Patent Office, to repeat Judge Michel's term. At least, thanks to Judge Michel and others, many more patent examiners have been hired since the "America Invents" law went into effect. Also, Congress no longer diverts the fees the Office collects in order to fund other government programs. Inroads are steadily being made in reducing the backlog of cases awaiting first "Office Action"—also increasing the quality of reviews.

But Congress and the Patent Office have long needed to take action to prevent patent trolls from terrorizing the inventor community. As the Cisco petition points out, the distinguishing feature of these bad guys is that they have no intention of manufacturing the patents they've acquired. Indeed, they're called "non-practicing entities—NPEs, or patent assertion entities—PAEs." (Count on the word mechanics among the legal profession to use euphemisms rather than calling these guys what they are, really, as Cisco calls them—"extortionists").

The door is wide open for the patent troll to challenge the validity of the original patent on seemingly any grounds. All he needs to do is to file a "Re-Examination" request with the Patent Trial and Appeals Board. "If," quoting from rules adopted under the "America Invents" law, "there is a showing that the challenge to the original patent will prevail," (even for one claim) then the patent troll will be able to make his case before the

Board, beginning his harassment of the owner of the original patent.

The Patent Office fees make it easy for the patent troll—and difficult for the owner—to make his case. He simply adds claims in his petition, the more claims the higher the fee—no problem for the typically well-financed troll, but a serious one for the typically not so well-financed owner, who must decide whether he wants to pay out the lawyer fees and related expenses in his defense before the Board. By now, the inevitable questions have come up for the owner—now the defendant. "Is my patent that good? Will my claims stand up? Should I take a chance, spending that much money in my defense? Is it all really worth it?"

That makes the original owner an easy mark for the troll, who will demand a relatively minor amount in license fees from the owner in consideration of his withdrawing his defense before the Board. The owner then keeps his patent, and the troll makes off with his bounty.

In the meanwhile, patent trolls have decided that small start-up companies are good hunting grounds. Colleen Chien, of the Santa Clara Law School, has made a recent study of the experiences of small start-up companies with new patents that have been approached by trolls (Santa Clara Univ. Legal Studies No. 09-12) . Unlike a Cisco, with significant cash reserves and other resources, the small start-up firm typically is cash poor, or at least lacking in other resources, so that many in her survey were virtually helpless, resulting, as Chien points out," in a significant operational impact for them...including delays in hiring, deferring milestones, suffering reduced valuations, even shutting down operations entirely," essentially giving up.

JOBS, JOBS, JOBS, was the slogan during this recent presidential election. It's well known that start-ups are "job incubators." The question: what if Google, with its 20 employees at start-up—now 30,000—had been forced to delay its milestones, deferred hiring, or shut down its operations entirely, or had given up altogether because of a patent troll's threats?

Or, if Apple, with its few hundred workers at the beginning, had experienced a similar threat? This gives rise to the related question: would the hundreds of thousands (millions?) now working for firms with important or related ties to Google or Apple—or any other start-up—be employed in some other organization? Obviously, the answer cannot even be guessed at. Would they have been employed at all?

"America Invents" was the most important, the boldest, patent legislation in years. Equally bold steps need to be taken now to deal with the scourge of patent trolls, emboldened by their continuing success, especially those who prey upon start-ups. For example:

Why not increase—significantly—the Re-Examination fees for the challenger, the troll, and maintain a low fee structure for the original owner in defense of his patent.

Or,

Since these guys lack anything resembling a significant "manufacturing record" what would be wrong with making the credible proof of such a record a condition precedent to the formal challenge to the original patent owner,

Or,

Having the applicant's request approved provided he agrees to manufacture/distribute the invention in question on a "best efforts" basis. And only then would the proceedings begin, with the burden of proof lying heavily on the challenger.

President Obama recently decided to take action against the troll. He has issued a number of executive orders calling for, among other things, disclosure of the true owners of patents, stiffer examination of patent applications, and other actions that would presumably serve to protect the victims of these predators. He called on Congress to pass laws that would make it tougher for the troll to sue. It remains to be seen whether these actions, in any meaningful way, will serve to thwart this piracy.

No matter what action is taken, there is urgent need to address this major issue now, taking into account our nation's still serious employment problem.

One can only speculate how much harm has been done to America's economy, to our social fabric, by a dysfunctional Patent Office and by an equally dysfunctional Congress. "America Invents" is a long overdue reform, but much more action needs to be taken, especially in addressing the harm done by patent trolls.

But a few examples, in addition to those pointed out earlier, may point to the magnitude of the problem.

America's historic leadership in manufacturing technology has long been displaced by Japan. According to Wikipedia, "While its population has increased less than 30 percent over twenty-five years, Japan's gross national product has increased thirtyfold; this growth has resulted in large part from rapid displacement of manual operations with innovative, high-speed, large-scale, continuously running. complex machines that process a growing number of miniaturized components."

Genome-sequencing, a major innovation in bioscience developed by American inventor/entrepreneur, Craig Venter, is now dominated by a Chinese company, Beijing Genomics Institute, which is acknowledged to be the largest genomic-research organization in the world, with thousands of workers. In a recent Science article, "BGI claimed to have more sequencing capacity than all U.S. labs combined."

America still leads the world in science and technology, seemingly in spite of the unwitting blocks and barriers that Congress allows to remain in place, but the question must be asked: how long will our dominance continue? Will it ultimately give way to the important strides that China and others are making?

leoncooper@verizon.net
P. O. Box 6030, Malibu, CA 90265

APPENDICES

Appendix A- Doctrine Of Equivalents in Patent Infringement

Appendix B – Patent Applications, By Year

Appendix C – Patent Office Fee Schedule

Appendix D – Cooper Patent

APPENDIX A

(The Doctrine Of Equivalents)

The Doctrine of Equivalents in Patent Infringement

Walter J. Blenko, Jr.

A patent contains several parts—a specification, usually one or more drawings, and always one or more claims. No matter how much a questioned machine, manufacture, composition of matter or process may look like the specification and drawings of a patent, it is only the claims of the patent which can be infringed. For that reason, if an issue of infringement arises, it becomes necessary to examine the claims of the patent in question.

The first step is to "read" each claim of the patent upon the accused structure or process. Every requirement of each claim must be considered to see if each thing set out in the claim also appears in the accused practice. If one or more things set forth in a claim is not present in the practice being reviewed, there is no infringement of that claim. On the other hand, if each thing which is set out in even one claim of the patent is present in the accused structure or process, then there is direct and literal infringement. When literal infringement is found, that is normally the end of the inquiry.

When the claims of a patent are read against an accused practice, they may be so close to identical that infringement is clear. Also, the accused practice may be so remote from the patent that there is no possibility of infringement. Very often, however, there are some differences, requiring further study. Sometimes, such differences are incorporated into a design after knowledge of a patent in an effort to avoid infringement. Then, the question presented is whether the design is sufficiently different from the patent to be held to be non- infringing. If the design is too close to the patent, it will infringe. If the design is remote enough, it will not infringe. The U.S. Supreme Court has stated: "One who seeks to pirate an invention, like one who seeks to pirate a copyrighted book or play, may be expected to introduce minor variations to conceal and shelter the piracy. Outright and forthright duplication is a dull and very rare form of infringement."

One reason that literal infringement of a patent is a dull form of infringement is that where the potential infringer knows of the patent and takes steps to avoid infringement by making changes from the exact thing which is disclosed and claimed in the patent. At the same time, the individual may copy as much of the patent as thought possible without becoming liable for infringement. When that happens, the issue raised is whether the accused structure or process is the "equivalent" of what is claimed in the patent. A similar problem can arise where a practice is adopted without knowledge of a patent, and the patent becomes known only after a business commitment has been made to follow the practice.

The rule of law for determining equivalency as laid down by the Supreme Court is quite simple: "If two devices do the same work in substantially the same way, and accomplish substantially the same result, they are the same, even though they differ in name, form, or shape." Despite the seeming simplicity of this rule, its application to a particular case is often complex. Determination of equivalency frequently involves conflicting opinions of experts and disputes as to scientific or engineering facts. The issue is often resolved by the testimony of expert witnesses and the decision may rest on which of the experts is the more believable. Things which are equivalent for one purpose may not be for other purposes. In one well- known case, the patent included claims for welding rods having a flux containing a major proportion of "alkaline earth metal silicate." The accused welding rods included a flux composed principally of "manganese silicate," which is not an alkaline earth metal silicate. Nevertheless, the accused welding rods were held to infringe because of testimony that manganese and magnesium were similar in many of their reactions and that they served the same purposes in fluxes.

Most patents are not issued with the claims originally filed. The claims of most patent applications are rejected, and the claims are amended with the inclusion of more detailed and restrictive language. If an examiner rejects the claims as unpatentable over the prior art and the claims are amended to read more narrowly to avoid the prior art, the patentee is barred from asserting the claims in the broader sense. Since he or she gave up the broader construction to obtain allowance of the claims, the patentee is not permitted to assert that the broader construction is the equivalent of the claim which was finally allowed. The process of rejection and amendment followed by allowance is shown by correspondence in the Patent and Trademark Office file. That file history is contained in a heavy paper jacket known as a "file wrapper." As a result, the rejection of a claim followed by a narrowing and more limiting amendment is known as a "file wrapper estoppel."

When a question of equivalency is under study, it is also necessary to know if the patent is a pioneer in a whole new field or if the patent shows only a narrow improvement of a subject that is well known. A pioneer patent will be given a much broader range of equivalents than one for a narrow improvement to existing technology.

It is not enough to rule out literal infringement. It is also necessary to determine equivalency. Equivalency can only be resolved by a careful analysis of the file history of the patent application. Such analysis should always be made when an accused structure or process resembles in any way a claim of a patent.

It is also possible for a device to be so far changed in principle from a patented device that it performs the same or a similar function in a substantially different way, even though it falls within the literal words of the claim. This situation can occur when the accused device is so far removed from the invention as disclosed in the patent that it is considered in law to be a different thing entirely. Even if the claims literally read on the thing which is accused, the claims are limited by construction to cover the invention which was disclosed in the patent and to exclude a thing which is different from the disclosed invention. This result is sometimes known as the "Reverse Doctrine of Equivalents."

Possible infringement of a patent claim must never be taken lightly. Issues of direct infringement, equivalency, file wrapper estoppel, and limitation of the claims to an invention as disclosed all must be considered before concluding that infringement does or does not exist.

Walter J. Blenko, Jr., is a senior partner in the law firm Eckert Seamans Cherin & Mellott, 600 Grant Street, 42nd Floor, Pittsburgh, PA 15219; telephone (412) 566-6000; fax (412) 566-6099; e-mail ARNIE@TEL-ERAMA.LM.COM.

Appendix A: The Doctrine of Equivalents in Patent Infringement from JOM, Volume 42, Issue 5, May 1990, P. 59. Copyright ©1990 by The Minerals, Metals & Materials Society. Reprinted with permission.

APPENDIX B

(Patents Applied/Patents Granted)

APPENDIX B

APPENDIX C

(Patent Office Fees)

APPENDIX C

 United States Patent and Trademark Office   FEES

Home | Site Index | Search | FAQ | Glossary | Guides | Contacts | eBusiness | eBiz alerts | News | Help

USPTO Fee Information > Current Fee Schedule

UNITED STATES PATENT AND TRADEMARK OFFICE
FEE SCHEDULE

Effective March 19, 2013 (Last Revised on June 3, 2013)

The fees subject to reduction upon establishment of small entity status (37 CFR 1.27) or micro entity status (37 CFR 1.29) are shown in separate columns. Payments from foreign countries must be payable and immediately negotiable in the United States for the full amount of the fee required. For additional information, please call the USPTO Contact Center at (571) 272-1000 or (800) 786-9199.

The $400/$200 non-electronic filing fee (fee codes 1090/2090 or 1690/2690) must be paid in addition to the filing, search and examination fees, in each original nonprovisional utility application filed in paper with the USPTO. The only way to avoid payment of the non-electronic filing fee is by filing your nonprovisional utility application via EFS-Web. The non-electronic filing fee does not apply to reissue, design, plant, or provisional applications.

Patent
Patent Application Filing Fees
Patent Search Fees
Patent Examination Fees
Patent Post-Allowance Fees
Patent Extension of Time Fees
Patent Maintenance Fees
Miscellaneous Patent Fees
Post Issuance Fees
Patent Trial and Appeal Fees
Patent Petition Fees
Patent Service Fees
Patent Enrollment Fees

Patent Cooperation Treaty
PCT Fees - National Stage
PCT Fees - International Stage
PCT Fees to Foreign Offices

General
Finance Service Fees
Computer Service Fees

Trademark
Trademark Processing Fees
Trademark Madrid Protocol Fees
Trademark International Application Fees
Trademark Service Fees
Fastener Quality Act Fees

Fee Code	37 CFR	Description	Fee	Small Entity Fee	Micro Entity Fee
Patent Application Filing Fees					
1011/2011/3011	1.16(a)	Basic filing fee - Utility	280.00	140.00	70.00
4011†	1.16(a)	Basic filing fee - Utility (electronic filing for small entities)		70.00	
1012/2012/3012	1.16(b)	Basic filing fee - Design	180.00	90.00	45.00
1017/2017/3017	1.16(b)	Basic filing fee - Design (CPA)	180.00	90.00	45.00
1013/2013/3013	1.16(c)	Basic filing fee - Plant	180.00	90.00	45.00
1005/2005/3005	1.16(d)	Provisional application filing fee	260.00	130.00	65.00
1014/2014/3014	1.16(e)	Basic filing fee - Reissue	280.00	140.00	70.00
1019/2019/3019	1.16(e)	Basic filing fee - Reissue (CPA)	280.00	140.00	70.00
1051/2051/3051	1.16(f)	Surcharge - Late filing fee, search fee, examination fee or oath or declaration	140.00	70.00	35.00
1052/2052/3052	1.16(g)	Surcharge - Late provisional filing fee or cover sheet	60.00	30.00	15.00
1201/2201/3201	1.16(h)	Independent claims in excess of three	420.00	210.00	105.00
1204/2204/3204	1.16(h)	Reissue independent claims in excess of three	420.00	210.00	105.00

74

Code	Rule	Description			
1202/2202/3202	1.16(i)	Claims in excess of 20	80.00	40.00	20.00
1205/2205/3205	1.16(i)	Reissue claims in excess of 20	80.00	40.00	20.00
1203/2203/3203	1.16(j)	Multiple dependent claim	780.00	390.00	195.00
1081/2081/3081	1.16(s)	Utility Application Size Fee - for each additional 50 sheets that exceeds 100 sheets	400.00	200.00	100.00
1082/2082/3082	1.16(s)	Design Application Size Fee - for each additional 50 sheets that exceeds 100 sheets	400.00	200.00	100.00
1083/2083/3083	1.16(s)	Plant Application Size Fee - for each additional 50 sheets that exceeds 100 sheets	400.00	200.00	100.00
1084/2084/3084	1.16(s)	Reissue Application Size Fee - for each additional 50 sheets that exceeds 100 sheets	400.00	200.00	100.00
1085/2085/3085	1.16(s)	Provisional Application Size Fee - for each additional 50 sheets that exceeds 100 sheets	400.00	200.00	100.00
1090/2090	1.16(t)	Non-electronic filing fee — Utility (additional fee for applications filed in paper)	400.00	200.00	200.00
1053/2053/3053	1.17(i)(1)	Processing fee, except in provisional applications	140.00	70.00	35.00

† The 4000 series fee code may be used via EFS-Web

Patent Search Fees 〇 Back to Top

1111/2111/3111	1.16(k)	Utility Search Fee	600.00	300.00	150.00
1112/2112/3112	1.16(l)	Design Search Fee	120.00	60.00	30.00
1113/2113/3113	1.16(m)	Plant Search Fee	380.00	190.00	95.00
1114/2114/3114	1.16(n)	Reissue Search Fee	600.00	300.00	150.00

Patent Examination Fees 〇 Back to Top

1311/2311/3311	1.16(o)	Utility Examination Fee	720.00	360.00	180.00
1312/2312/3312	1.16(p)	Design Examination Fee	460.00	230.00	115.00
1313/2313/3313	1.16(q)	Plant Examination Fee	580.00	290.00	145.00
1314/2314/3314	1.16(r)	Reissue Examination Fee	2,160.00	1,080.00	540.00

Patent Post-Allowance Fees 〇 Back to Top

1501/2501/3501	1.18(a)(2)	Utility issue fee	1,780.00	890.00	445.00
1511/2511/3511	1.18(a)(2)	Reissue issue fee	1,780.00	890.00	445.00
1502/2502/3502	1.18(b)(2)	Design issue fee	1,020.00	510.00	255.00
1503/2503/3503	1.18(c)(2)	Plant issue fee	1,400.00	700.00	350.00
1504	1.18(d)(2)	Publication fee for early, voluntary, or normal publication	300.00	300.00	300.00
1505	1.18(d)(3)	Publication fee for republication	300.00	300.00	300.00

Patent Extension of Time Fees 〇 Back to Top

1251/2251/3251	1.17(a)(1)	Extension for response within first month	200.00	100.00	50.00
1252/2252/3252	1.17(a)(2)	Extension for response within second month	600.00	300.00	150.00
1253/2253/3253	1.17(a)(3)	Extension for response within third month	1,400.00	700.00	350.00
1254/2254/3254	1.17(a)(4)	Extension for response within fourth month	2,200.00	1,100.00	550.00
1255/2255/3255	1.17(a)(5)	Extension for response within fifth month	3,000.00	1,500.00	750.00

Patent Maintenance Fees
Back to Top

1551/2551/3551	1.20(e)	Due at 3.5 years	1,600.00	800.00	400.00
1552/2552/3552	1.20(f)	Due at 7.5 years	3,600.00	1,800.00	900.00
1553/2553/3553	1.20(g)	Due at 11.5 years	7,400.00	3,700.00	1,850.00
1554/2554/3554	1.20(h)	Surcharge - 3.5 year - Late payment within 6 months	160.00	80.00	40.00
1555/2555/3555	1.20(h)	Surcharge - 7.5 year - Late payment within 6 months	160.00	80.00	40.00
1556/2556/3556	1.20(h)	Surcharge - 11.5 year - Late payment within 6 months	160.00	80.00	40.00
1557/2557/3557	1.20(i)(1)	Surcharge after expiration - Late payment is unavoidable	700.00	350.00	175.00
1558/2558/3558	1.20(i)(2)	Surcharge after expiration - Late payment is unintentional	1,640.00	820.00	410.00

Miscellaneous Patent Fees
Back to Top

1817/2817/3817	1.17(c)	Request for prioritized examination	4,000.00	2,000.00	1,000.00
1819/2819/3819	1.17(d)	Correction of inventorship after first action on merits	600.00	300.00	150.00
1801/2801/3801	1.17(e)(1)	Request for continued examination (RCE) - 1st request (see 37 CFR 1.114)	1,200.00	600.00	300.00
1820/2820/3820	1.17(e)(2)	Request for continued examination (RCE) - 2nd and subseqent request (see 37 CFR 1.114)	1,700.00	850.00	425.00
1808	1.17(i)(2)	Other publication processing fee	130.00	130.00	130.00
1803	1.17(i)(2)	Request for voluntary publication or republication	130.00	130.00	130.00
1802/2802/3802	1.17(k)	Request for expedited examination of a design application	900.00	450.00	225.00
1806/2806/3806	1.17(p)	Submission of an Information Disclosure Statement	180.00	90.00	45.00
1818/2818	1.17(p)	Document fee for third-party submissions (see 37 CFR 1.290(f))	180.00	90.00	
1807	1.17(q)	Processing fee for provisional applications	50.00	50.00	50.00
1809/2809/3809	1.17(r)	Filing a submission after final rejection (see 37 CFR 1.129(a))	840.00	420.00	210.00
1810/2810/3810	1.17(s)	For each additional invention to be examined (see 37 CFR 1.129(b))	840.00	420.00	210.00

Post Issuance Fees
Back to Top

1811	1.20(a)	Certificate of correction	100.00	100.00	100.00
1816	1.20(b)	Processing fee for correcting inventorship in a patent	130.00	130.00	130.00
1812/2812/3812	1.20(c)(1)	Request for ex parte reexamination	12,000.00	6,000.00	3,000.00*
1821/2821/3821	1.20(c)(3)	Reexamination independent claims in excess of three and also in excess of the number of such claims in the patent under reexamination	420.00	210.00	105.00
1822/2822/3822	1.20(c)(4)	Reexamination claims in excess of 20 and also in excess of the number of claims in the patent under reexamination	80.00	40.00	20.00
1814	1.20(d)	Statutory disclaimer, including terminal disclaimer	160.00	160.00	160.00
1826/2826/3826	1.20(k)(1)	Request for supplemental examination	4,400.00	2,200.00	1,100.00
1827/2827/3827	1.20(k)(2)	Reexamination ordered as a result of supplemental examination	12,100.00	6,050.00	3,025.00
1828/2828/3828	1.20(k)(3)(i)	Supplemental Examination Document Size Fee - for nonpatent document having between 21 and 50	180.00	90.00	45.00

Code	CFR	Description			
1828/2828/3828	1.20(k)(3)(i)	Supplemental Examination Document Size Fee - for nonpatent document having between 21 and 50 sheets	180.00	90.00	45.00
1829/2829/3829	1.20(k)(3)(ii)	Supplemental Examination Document Size Fee - for each additional 50 sheets or a fraction thereof in a nonpatent document	280.00	140.00	70.00

* Third-party filers are not eligible for the micro entity fee.

Patent Trial and Appeal Fees 🔝 Back to Top

Code	CFR	Description			
1405	41.20(a)	Petitions to the Chief Administrative Patent Judge under 37 CFR 41.3	400.00	400.00	400.00
1401/2401/3401	41.20(b)(1)	Notice of appeal	800.00	400.00	200.00*
n/a	41.20(b)(2)(i)	Filing a brief in support of an appeal	0.00	0.00	0.00
1404/2404/3404	41.20(b)(2)(ii)	Filing a brief in support of an appeal in an inter partes reexamination proceeding	2,000.00	1,000.00	500.00*
1403/2403/3403	41.20(b)(3)	Request for oral hearing	1,300.00	650.00	325.00*
1413/2413/3413	41.20(b)(4)	Forwarding an appeal in an application or ex parte reexamination proceeding to the Board	2,000.00	1,000.00	500.00*
1406	42.15(a)(1)	Inter partes review request fee - Up to 20 claims	9,000.00	9,000.00	9,000.00
1414	42.15(a)(2)	Inter partes review post-institution fee - Up to 15 claims	14,000.00	14,000.00	14,000.00
1407	42.15(a)(3)	Inter partes review request of each claim in excess of 20	200.00	200.00	200.00
1415	42.15(a)(4)	Inter partes post-institution request of each claim in excess of 15	400.00	400.00	400.00
1408	42.15(b)(1)	Post-grant or covered business method review request fee - Up to 20 claims	12,000.00	12,000.00	12,000.00
1416	42.15(b)(2)	Post-grant or covered business method review post-institution fee - Up to 15 claims	18,000.00	18,000.00	18,000.00
1409	42.15(b)(3)	Post-grant or covered business method review request of each claim in excess of 20	250.00	250.00	250.00
1417	42.15(b)(4)	Post-grant or covered business method review post-institution request of each claim in excess of 15	550.00	550.00	550.00
1412	42.15(c)(1)	Petition for a derivation proceeding	400.00	400.00	400.00
1411	42.15(d)	Request to make a settlement agreement available and other requests filed in a patent trial proceeding	400.00	400.00	400.00

* Third-party filers are not eligible for the micro entity fee.

Patent Petition Fees 🔝 Back to Top

Code	CFR	Description			
1462/2462/3462	1.17(f)	Petitions requiring the petition fee set forth in 37 CFR 1.17(f) (Group I)	400.00	200.00	100.00
1463/2463/3463	1.17(g)	Petitions requiring the petition fee set forth in 37 CFR 1.17(g) (Group II)	200.00	100.00	50.00
1464/2464/3464	1.17(h)	Petitions requiring the petition fee set forth in 37 CFR 1.17(h) (Group III)	140.00	70.00	35.00

1452/2452/3452	1.17(l)	Petition to revive unavoidably abandoned application	640.00	320.00	160.00
1453/2453/3453	1.17(m)	Petition to revive unintentionally abandoned application	1,900.00	950.00	475.00
1454/2454/3454	1.17(t)	Acceptance of an unintentionally delayed claim for priority, or for filing a request for the restoration of the right of priority	1,420.00	710.00	355.00
1455	1.18(e)	Filing an application for patent term adjustment	200.00	200.00	200.00
1456	1.18(f)	Request for reinstatement of term reduced	400.00	400.00	400.00
1824/2824/3824	1.20(c)(6)	Petitions in a reexamination proceeding, except for those specifically enumerated in 37 CFR 1.550(i) and 1.937(d)	1,940.00	970.00	485.00*
1457	1.20(j)(1)	Extension of term of patent	1,120.00	1,120.00	1,120.00
1458	1.20(j)(2)	Initial application for interim extension (see 37 CFR 1.790)	420.00	420.00	420.00
1459	1.20(j)(3)	Subsequent application for interim extension (see 37 CFR 1.790)	220.00	220.00	220.00

* Third-party filers are not eligible for the micro entity fee.

PCT Fees - National Stage   🎧 Back to Top

1631/2631/3631	1.492(a)	Basic National Stage Fee	280.00	140.00	70.00
n/a	1.492(b)(1)	National Stage Search Fee - U.S. was the ISA or IPEA and all claims satisfy PCT Article 33(1)-(4)	0.00	0.00	0.00
1641/2641/3641	1.492(b)(2)	National Stage Search Fee - U.S. was the ISA	120.00	60.00	30.00
1642/2642/3642	1.492(b)(3)	National Stage Search Fee - search report prepared and provided to USPTO	480.00	240.00	120.00
1632/2632/3632	1.492(b)(4)	National Stage Search Fee - all other situations	600.00	300.00	150.00
n/a	1.492(c)(1)	National Stage Examination Fee - U.S. was the ISA or IPEA and all claims satisfy PCT Article 33(1)-(4)	0.00	0.00	0.00
1633/2633/3633	1.492(c)(2)	National Stage Examination Fee - all other situations	720.00	360.00	180.00
1614/2614/3614	1.492(d)	Claims - extra independent (over three)	420.00	210.00	105.00
1615/2615/3615	1.492(e)	Claims - extra total (over 20)	80.00	40.00	20.00
1616/2616/3616	1.492(f)	Claims - multiple dependent	780.00	390.00	195.00
1617/2617/3617	1.492(h)	Search fee, examination fee or oath or declaration after the date of commencement of the national stage	140.00	70.00	35.00
1618/2618/3618	1.492(i)	English translation after thirty months from priority date	140.00	70.00	35.00
1681/2681/3681	1.492(j)	National Stage Application Size Fee - for each additional 50 sheets that exceeds 100 sheets	400.00	200.00	100.00

PCT Fees - International Stage   🎧 Back to Top

1601	1.445(a)(1)(i)(B)	Transmittal fee	240.00	240.00	240.00
1690/2690	1.445(a)(1)(ii)	Non-electronic filing fee (additional fee for applications filed in paper)	400.00	200.00	200.00
1602	1.445(a)(2)(ii)	Search fee - regardless of whether there is a corresponding application (see 35 U.S.C. 361(d) and PCT Rule 16)	2,080.00	2,080.00	2,080.00
1604	1.445(a)(3)(ii)	Supplemental search fee when required, per additional invention	2,080.00	2,080.00	2,080.00

1621	1.445(a)(4)(ii)	Transmitting application to Intl. Bureau to act as receiving office	240.00	240.00	240.00
1605	1.482(a)(1)(i)(B)	Preliminary examination fee - U.S. was the ISA	600.00	600.00	600.00
1606	1.482(a)(1)(ii)(B)	Preliminary examination fee - U.S. was not the ISA	750.00	750.00	750.00
1607	1.482(a)(2)(ii)	Supplemental examination fee per additional invention	600.00	600.00	600.00
1619		Late payment fee	variable	variable	variable
PCT Fees to Foreign Offices *					🎧 Back to Top
1701		International filing fee (first 30 pages - filed in paper with PCT EASY zip file or electronically without PCT EASY zip file)	1,312.00	1,312.00	1,312.00
1710		International filing fee (first 30 pages) - filed electronically with PCT Easy zip file	1,206.00	1,206.00	1,206.00
1702		International filing fee (first 30 pages)	1,419.00	1,419.00	1,419.00
1703		Supplemental fee (for each page over 30)	16.00	16.00	16.00
1704		International search (EPO)	2,419.00	2,419.00	2,419.00
1712		International search (IPAU)	2,282.00	2,282.00	2,282.00
1709		International search (KIPO)	1,167.00	1,167.00	1,167.00
1714		International search (Rospatent)	217.00	217.00	217.00
1705		Handling fee	213.00	213.00	213.00
1706		Handling Fee - 90% reduction, if applicants meets criteria specified at: http://www.wipo.int/pct/en/fees/fee_reduction.pdf	21.30	21.30	21.30

** PCT Fees to Foreign Offices subject to periodic change due to fluctuations in exchange rate.

Patent Service Fees					🎧 Back to Top
8001	1.19(a)(1)	Printed copy of patent w/o color, delivery by USPS, USPTO Box, or electronic means	3.00	3.00	3.00
8005	1.19(a)(1)	Patent Application Publication (PAP)	3.00	3.00	3.00
8003	1.19(a)(2)	Printed copy of plant patent in color	15.00	15.00	15.00
8004	1.19(a)(3)	Color copy of patent (other than plant patent) containing a color drawing	25.00	25.00	25.00
8007	1.19(b)(1)(i)(A), (ii)(A), and (iii)(A)	Copy of patent application as filed	20.00	20.00	20.00
8008	1.19(b)(1)(i)(B)	Copy of patent-related file wrapper and contents of 400 or fewer pages, if provided on paper	200.00	200.00	200.00
8009	1.19(b)(1)(i)(C)	Additional fee for each additional 100 pages of patent-related file wrapper and (paper) contents, or portion thereof	40.00	40.00	40.00
8010	1.19(b)(1)(i)(D)	Individual application documents, other than application as filed, per document	25.00	25.00	25.00
8011	1.19(b)(1)(ii)(B) and (iii)(B)	Copy of patent-related file wrapper and contents if provided electronically or on a physical electronic medium as specified in 1.19(b)(1)(ii)	55.00	55.00	55.00
8012	1.19(b)(1)(ii)(C)	Additional fee for each continuing physical electronic medium in single order of 1.19(b)(1)(ii)(B)	15.00	15.00	15.00
8041	1.19(b)(2)(i)(A)	Copy of patent-related file wrapper contents that were submitted and are stored on compact disk or other electronic form (e.g., compact disks stored in Artifact	55.00	55.00	55.00

8042	1.19(b)(2)(i)(B)	Additional fee for each continuing copy of patent-related file wrapper contents as specified in 1.19(b)(2)(i)(A)	15.00	15.00	15.00
8043	1.19(b)(2)(ii)	Copy of patent-related file wrapper contents that were submitted and are stored on compact disk, or other electronic form, other than as available in 1.19(b)(1); if provided electronically other than on a physical electronic medium, per order	55.00	55.00	55.00
8013	1.19(b)(3)	Copy of office records, except copies of applications as filed	25.00	25.00	25.00
8014	1.19(b)(4)	For assignment records, abstract of title and certification, per patent	25.00	25.00	25.00
8904	1.19(c)	Library service	50.00	50.00	50.00
8016	1.19(e)	Uncertified statement re status of maintenance fee payments	10.00	10.00	10.00
8017	1.19(f)	Copy of non-U.S. document	25.00	25.00	25.00
8050	1.19(g)	Petitions for documents in form other than that provided by this part, or in form other than that generally provided by Director, to be decided in accordance with merits	AT COST	AT COST	AT COST
8020	1.21(e)	International type search report	40.00	40.00	40.00
8902	1.21(g)	Self-service copy charge, per page	0.25	0.25	0.25
8021	1.21(h)(2)	Recording each patent assignment, agreement or other paper, per property	40.00	40.00	40.00
8022	1.21(i)	Publication in Official Gazette	25.00	25.00	25.00
8023	1.21(j)	Labor charges for services, per hour or fraction thereof	40.00	40.00	40.00
8024	1.21(k)	Unspecified other services, excluding labor	AT COST	AT COST	AT COST
8026	1.21(n)	Handling fee for incomplete or improper application	130.00	130.00	130.00

Patent Enrollment Fees				⌂ Back to Top	
9001	1.21(a)(1)(i)	Application fee (non-refundable)	40.00		
9010	1.21(a)(1)(ii)(A)	For test administration by commercial entity	200.00		
9011	1.21(a)(1)(ii)(B)	For test administration by the USPTO	450.00		
9003	1.21(a)(2)	Registration to practice or grant of limited recognition under §11.9(b) or (c)	100.00		
9025	1.21(a)(2)	Registration to practice for change of practitioner type	100.00		
9005	1.21(a)(4)	Certificate of good standing as an attorney or agent	10.00		
9006	1.21(a)(4)(i)	Certificate of good standing as an attorney or agent, suitable for framing	20.00		
9012	1.21(a)(5)(i)	Review of decision by the Director of Enrollment and Discipline under §11.2(c)	130.00		
9013	1.21(a)(5)(ii)	Review of decision of the Director of Enrollment and Discipline under §11.2(d)	130.00		
9015	1.21(a)(7)(i)	Annual fee for registered attorney or agent in active status	120.00		
9016	1.21(a)(7)(ii)	Annual fee for registered attorney or agent in voluntary inactive status	25.00		
9017	1.21(a)(7)(iii)	Requesting restoration to active status from voluntary inactive status	50.00		

9018	1.21(a)(7)(iv)	Balance of annual fee due upon restoration to active status from voluntary inactive status	100.00	
9019	1.21(a)(8)	Annual fee for individual granted limited recognition	120.00	
9020	1.21(a)(9)(i)	Delinquency fee for annual fee	50.00	
9004	1.21(a)(9)(ii)	Reinstatement to practice	100.00	
9014	1.21(a)(10)	Application fee for person disciplined, convicted of a felony or certain misdemeanors under §11.7(h)	1,600.00	
9024	1.21(k)	Unspecified other services, excluding labor	AT COST	

Finance Service Fees   🔼 Back to Top

9201	1.21(b)(1) or 2.6(b)(13)(i)	Establish deposit account	10.00	
9202	1.21(b)(2), (b)(3) or 2.6(b)(13)(ii)	Service charge for below minimum balance	25.00	
9101	1.21(m) or 2.6(b)(12)	Processing each payment refused or charged back	50.00	

Computer Service Fees   🔼 Back to Top

8031		Computer records	AT COST	

Trademark Processing Fees ***   🔼 Back to Top

6001	2.6(a)(1)(i)	Application for registration, per international class (paper filing)	375.00	
7001	2.6(a)(1)(ii)	Application for registration, per international class (electronic filing, TEAS application)	325.00	
7007	2.6(a)(1)(iii)	Application for registration, per international class (electronic filing, TEAS Plus application)	275.00	
6002/7002	2.6(a)(2)	Filing an Amendment to Allege Use under §1(c), per class	100.00	
6003/7003	2.6(a)(3)	Filing a Statement of Use under §1(d)(1), per class	100.00	
6004/7004	2.6(a)(4)	Filing a Request for a Six-month Extension of Time for Filing a Statement of Use under §1(d)(1), per class	150.00	
6005/7005	2.6(a)(15)	Petitions to the Director	100.00	
6006/7006	2.6(a)(19)	Dividing an application, per new application (file wrapper) created	100.00	
6008/7008	2.6(a)(1)(iv)	Additional fee for application that doesn't meet TEAS Plus filing requirements, per class	50.00	
6201/7201	2.6(a)(5)	Application for renewal under §9, per class	400.00	
6203/7203	2.6(a)(6)	Additional fee for filing renewal application during grace period, per class	100.00	
6204/7204	2.6(a)(21)	Correcting a deficiency in a renewal application	100.00	
6205/7205	2.6(a)(12)	Filing §8 affidavit, per class	100.00	
6206/7206	2.6(a)(14)	Additional fee for filing §8 affidavit during grace period, per class	100.00	
6207/7207	2.6(a)(20)	Correcting a deficiency in a §8 affidavit	100.00	
6208/7208	2.6(a)(13)	Filing §15 affidavit, per class	200.00	
6210/7210	2.6(a)(7)	Publication of mark under §12(c), per class	100.00	
6211/7211	2.6(a)(8)	Issuing new certificate of registration	100.00	
6212/7212	2.6(a)(9)	Certificate of correction, registrant's error	100.00	

6213/7213	2.6(a)(10)	Filing disclaimer to registration	100.00	
6214/7214	2.6(a)(11)	Filing amendment to registration	100.00	
6215/7215	7.37	Filing §71 affidavit, per class	100.00	
6216/7216	7.37	Filing §71 affidavit grace period, per class	100.00	
6401/7401	2.6(a)(16)	Petition for cancellation, per class	300.00	
6402/7402	2.6(a)(17)	Notice of opposition, per class	300.00	
6403/7403	2.6(a)(18)	Ex parte appeal, per class	100.00	

Trademark Madrid Protocol Fees ***   ⬆ Back to Top

6901/7901	7.6(a)(1)	Certifying an International application based on single application or registration, per class	100.00	
6902/7902	7.6(a)(2)	Certifying an International application based on more than one basic application or registration, per class	150.00	
6903/7903	7.6(a)(3)	Transmitting a Request to Record an Assignment or restriction under 7.23 or 7.24	100.00	
6904/7904	7.6(a)(4)	Filing a Notice of Replacement, per class	100.00	
6905/7905	7.6(a)(5)	Filing an affidavit under 71 of the Act, per class	100.00	
6906/7906	7.6(a)(6)	Surcharge for filing affidavit under 71 of the Act during grace period, per class	100.00	
6907/7907	7.6(a)(7)	Transmitting a subsequent designation	100.00	
6908/7908	7.6(a)(8)	Correcting a deficiency in an affidavit under 71 of the Act	100.00	

Trademark International Application Fees ***   ⬆ Back to Top

7951	7.7(1)	International application fee	Reference CFR 7.7 for payment of fees to International Bureau (IB) and IB calculator at: http://www.wipo.int/madrid/en.
7952	7.14(c)	Correcting irregularities in an International application	
7953	7.21	Subsequent designation fee	
7954	7.23	Recording of an assignment of an international registration under 7.23	

*** The 7000 series fee code (e.g., 7001, 7002, etc.) is used for electronic filing via **TEAS**.

Trademark Service Fees   ⬆ Back to Top

8501	2.6(b)(1)	Printed copy of registered mark, delivery by USPS, USPTO Box, or electronic means	3.00	
8503	2.6(b)(4)(i)	Certified copy of registered mark, with title and/or status, regular service	15.00	
8504	2.6(b)(4)(ii)	Certified copy of registered mark, with title and/or status, expedited local service	30.00	
8507	2.6(b)(2)	Certified copy of trademark application as filed	15.00	
8508	2.6(b)(3)	Certified or uncertified copy of trademark-related file wrapper and contents	50.00	
8513	2.6(b)(5)	Certified or uncertified copy of trademark document, unless otherwise provided	25.00	
8514	2.6(b)(7)	For assignment records, abstracts of title and certification per registration	25.00	
8902	2.6(b)(9)	Self-service copy charge, per page	0.25	
8521	2.6(b)(6)	Recording trademark assignment, agreement or other paper, first mark per document	40.00	

8522	2.6(b)(6)	For second and subsequent marks in the same document	25.00		
8523	2.6(b)(10)	Labor charges for services, per hour or fraction thereof	40.00		
8524	2.6(b)(11)	Unspecified other services, excluding labor	AT COST		
Fastener Quality Act Fees					🎧 Back to Top
6991	2.7(a)	Recordal application fee	20.00		
6992	2.7(b)	Renewal application fee	20.00		
6993	2.7(c)	Late fee for renewal application	20.00		
6994	2.7(a)	Application fee for reactivation of insignia, per request	20.00		

APPENDIX D

(Cooper Patent)

APPENDIX D

Patents

[Find prior art]　[Discuss this patent]　[Read this patent]　[Download PDF]　⚙ ▾

Smoke detector tester
William H. Haines et al

› Overview
Abstract
Drawings
Description
Claims

[Go]

Patent number: 4301674
Filing date: Jan 14, 1980
Issue date: Nov 24, 1981

A smoke detector tester is disclosed herein having a hand held aerosol container for holding under pressure a quantity of spray emitted from the container by a finger valve in the form of an aerosol cloud of known particle distribution within the sensing area of a smoke detector undergoing test. The particle size distribution within the aerosol cloud simulates the aerosol particle sizes that are emitted during the early stages of the process of combustion. The tester thus provides a true functional test of the smoke detector's sensing ability.

Inventors: William H. Haines, Leon C. Cooper
Current U.S. Classification: 73/1.06; 222/4; 516/2
International Classification: G01M 1900

View patent at USPTO
Search USPTO Assignment Database

Citations

Cited Patent	Filing date	Issue date	Original Assignee	Title
US3469723	May 4, 1967	Mar 19, 1970		ATMOSPHERIC POLLUTION CONTROL
US3696401	Nov 13, 1970	Sep 6, 1972		APPARATUS FOR CHECKING OPERATION OF
US3729979	May 28, 1971	May 1, 1973		COMBUSTION PRODUCTS GENERATING AND METERING DEVICE
US3985866	Jan 13, 1975	Oct 12, 1976	American Cyanamid Company	Asthma treatment by inhalation of micronized N,N-diethyl-4-methyl-1-piperazinecarboxamide pamoate

Referenced by

Citing Patent	Filing date	Issue date	Original Assignee	Title
US4716985	Oct 30, 1985	Dec 29, 1987	L'Air Liquide, Societe Anonyme pour l'Etude et l'Exploitation des Procedes George Claude	Composition for checking the functioning of fire detection installations and application to various types of detectors
US4990290	Apr 18, 1990	Feb 5, 1991		Diffusion fogger
US5057243	Jun 8, 1988	Oct 15, 1991	Pro Efx, Inc.	Aerosol diffusion fogger
US5060503	Feb 8, 1990	Oct 29, 1991	Bacharach, Inc.	Test kit for gas detectors
US5076996	Jul 2, 1990	Dec 31, 1991	John J. McSheffrey Kevin L. McSheffrey	Composition and method for testing smoke detectors
US5139690	Jun 27, 1990	Aug 18, 1992		Spray formulation for the testing of smoke detectors
US5240648	Feb 14, 1992	Aug 31, 1993		Compact fogger
US5309148	Dec 16, 1992	May 3, 1994		Apparatus and method for testing smoke detector operation
US5361623	Sep 30, 1992	Nov 6, 1994	Leon Cooper William H. Haines	Delivery system for smoke detector testing spray formulation
US5670946	Feb 23, 1996	Sep 23, 1997	No Climb Products Limited	Smoke detector sensitivity testing apparatus
US5785891	Sep 12, 1996	Jul 28, 1998	Leon Cooper	Spray formulation for the testing of smoke detectors
US6196399	Mar 9, 2000	Mar 8, 2001		Smoke detector test device and

86

United States Patent [19]

Haines et al.

[11] **4,301,674**

[45] **Nov. 24, 1981**

[54] SMOKE DETECTOR TESTER

[76] Inventors: William H. Haines, 5240 Topeka Dr., Tarzana, Calif. 91356; Leon C. Cooper, 31316 Via Colinas, Westlake Village, Calif. 91361

[21] Appl. No.: 111,826

[22] Filed: Jan. 14, 1980

[51] Int. Cl.³ G01M 19/00

[52] U.S. Cl. .... 73/1 G; 222/4; 252/305

[58] Field of Search .... 73/1 G; 222/4; 252/305

[56] References Cited

U.S. PATENT DOCUMENTS

3,499,723	3/1970	Hamilton et al.	252/305
3,693,401	9/1972	Purt et al.	73/1 G
3,729,979	5/1973	Wiberg	73/1 G
3,985,868	10/1976	Cory, Jr. et al.	222/4

Primary Examiner—S. Clement Swisher

[57] **ABSTRACT**

A smoke detector tester is disclosed herein having a hand held aerosol container for holding under pressure a quantity of spray emitted from the container by a finger valve in the form of an aerosol cloud of known particle distribution within the sensing area of a smoke detector undergoing test. The particle size distribution within the aerosol cloud simulates the aerosol particle sizes that are emitted during the early stages of the process of combustion. The tester thus provides a true functional test of the smoke detector's sensing ability.

3 Claims, 4 Drawing Figures

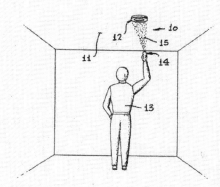

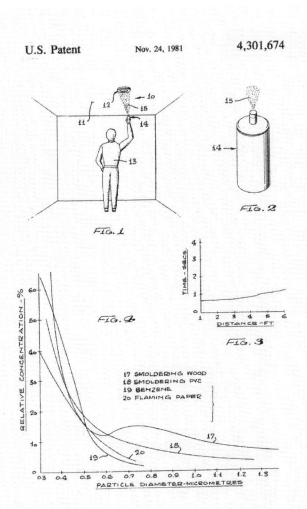

U.S. Patent   Nov. 24, 1981   4,301,674

FIG. 1

FIG. 2

FIG. 4

FIG. 3

17 SMOLDERING WOOD
18 SMOLDERING PVC
19 BENZENE
20 FLAMING PAPER

4,301,674

1

SMOKE DETECTOR TESTER

BACKGROUND OF THE INVENTION

1. Field of the Invention

This invention relates to apparatus for testing smoke detectors of both the ionization and the photoelectric type and more particularly to a novel such tester which involves actual simulation of fire conditions.

2. Brief Description of the Prior Art

There are two basic types of smoke detectors. One is the ionization detector, which senses changes in the conductivity of the air in a measuring chamber or chambers under the influence of radio-active radiation. The other is the photoelectric detector which senses the scattering of light in a measuring chamber. Both devices respond to the presence of particulate matter, or particles of combustion produced by thermal decomposition. The great majority of these particles, or aerosols, are invisible (less than 0.3 micrometers in size) but the larger ones are visible in the form of smoke (0.3 micrometers to 10.0 micrometers).

Invisible aerosol is the earliest appearing fire signature noted to date. Heating of materials during the pre-ignition stage of a fire produces submicron particles ranging in size from 5×10^{-4} to 1×10^{-3} micrometers. These particles are generated at temperatures well below ignition temperatures.

As heating of a material progresses toward the ignition temperature, the concentration of visible aerosol increases to the point where larger particles are formed by coagulation. As this process continues, the particle size distribution becomes log normal with the most frequent sizes in the range between 0.1 and 1.2 micrometers. This is the size range to which both ionization and photoelectric devices will respond, the photoelectric device reacting to sizes 0.3 micrometers and above and the ionization to the entire range and smaller. The smaller particles, less than 0.1 micrometers disappear either by coagulation or by evaporation, and the larger particles, greater than 1.0 micrometer, are lost through the processes of sedimentation following Stokes' Law. Aerosols in this size range are remarkably stable and contain particles in both the visible and invisible aerosol signature range. This "ageing" of aerosols has been reported by Van Luik and Scheidweiler.

Smoke detectors are a widely used home safety device which are designed to protect lives and to reduce property damage by giving a warning of fire when the fire is in its earliest stage in the home. If the home is equipped with one or more smoke detectors, the alarm, typically a loud horn, given by a properly functioning smoke detector can immediately alert the home occupants, giving them the time they need to safely exit the building. How fast the detector responds is of particular importance because the time interval between the warning alarm and the spread of the fire through the household is the critical factor. A few minutes is often the difference between survival and death in the typical home fire.

Detectors may fail to alarm as required because their sensitivities have been altered over a period of time by dust, grease, corrosive fumes, moisture or by other contaminants in the area in which the detector is located. Electronic component failures are also known to occur. Aging, as well, is a factor contributing to malfunctions for these reasons. Detector manufacturers typically provide for "testing" the device by means of

2

pressing a test button or by pulling a switch which is located on the housing of the unit. Alternatively, some (older) models can only be tested by blowing smoke in the direction of the detector, i.e., through smoke derived from a cigarette, cigar, match, candle, paper, rope, etc.

There are major disadvantages in the conventional testing of smoke detectors. Recommended placement of the smoke detector is on the ceiling or high-up on the wall. A person of average height must stand on a chair or on some elevation in order to press the button, pull the switch, or blow the smoke, thus risking physical harm, which is a serious matter for older people. At best, blowing smoke is a clumsy and primitive method for testing such devices.

The only true test for a smoke detector is one that involves creating the particular matter (aerosols) which simulates the advance or early products of combustion. Underwriters Laboratories, Inc. publishes a Standard for Safety for Single and Multiple Station Smoke Detectors (UL 217) which specifies the detection levels for an approved smoke detector. Compliance is determined by empirical tests involving, among others, flaming paper, gasoline and smoldering wood. A detector must function satisfactorily in all such tests. The levels of performance required are minimum standards, basically calling for any detector to operate an alarm when exposed to particle sizes in the range of 0.1 micrometers to 1.2 micrometers. See FIG. 4. So it is important that in testing smoke detectors the test should provide only those detectable elements within the prescribed minimum levels for an approved detector. It would be improper and ineffective to employ an "overkill" type test, such as used by Gustav Part, et al (U.S. Pat. No. 3,693,401) wherein a housing encloses the smoke detector, creating an artificial environment into which aerosols of indeterminate size are introduced in a gross quantity. Part's artificial environment, of course, precludes normal air flow conditions and the aerosols introduced are not limited as to quantity or appropriate size.

SUMMARY OF THE INVENTION

The problem and difficulties encountered with conventional testing of smoke detectors are obviated by the present invention which performs satisfactorily on both types of smoke detectors, ionization and photoelectric, because its chemical formulation produces a distribution of particle sizes which simulates the full range of fire conditions.

The following laboratory tests were conducted to determine the aerosol sizes of the invention:

TEST I

The test aerosol was injected into a smoke detector test tunnel to measure the apparent geometric diameter while the aerosol was being circulated in the tunnel. This tunnel is similar to the tunnels used by the Underwriters Laboratories for smoke detectors.

The tunnel is equipped with a ½meter light beam which is used to measure the density of the smoke in the tunnel using the extinction principle. The tunnel is also equipped with a measuring ionization chamber which is used to measure smoke density using the ionization chamber principle. Through the use of suitable mathematics, it is possible to determine the appropriate size of the smoke particles in the tunnel from the readings

4,301,674

3

obtained with the light beam and the measuring ionization chamber. Scheidweiler has described this procedure in the May 1976 issue of Fire Technology.

Using this technique, it was found that the Smoke Detector Tester's aerosol had a mean geometric diameter of 1.1 micrometer.

TEST II

In tests using a TechEcology Model 200 particle counter and a Tracor Northern Model TN-1705 multichannel pulse height analyzer, a mean geometric diameter of 0.7 micrometers (calculated from the mass median diameter) was obtained.

The measurements obtained in both tests indicate the suitability of the invention's particle sizes for the purpose intended.

The formulation is responsive to the common observation that ionization devices are more sensitive to smoke from flaming combustion, i.e., smaller particles, and that photoelectric devices are more sensitive to smoke from a smoldering source, i.e., larger particles. Specifically, a hand held aerosol container includes a finger operated dispensing means for selectively releasing a quantity of particles into the detector's sensing area. The particle size is small enough, in the case of an ionization device, to cause a voltage drop inside the ionization chamber . . . and large enough, in the case of a photoelectric device, to cause light scattering inside the detection apparatus; either condition will cause the alarm to sound, resulting in a true functional test.

Therefore, it is among the primary objects of the present invention to provide a novel smoke detector tester which may be readily employed to test the total sensing capabilities of the detector as well as the electrical circuitry connected therewith.

BRIEF DESCRIPTION OF THE DRAWINGS

The features of the present invention which are believed to be novel are set forth with particularity in the appended claims. The present invention, both as to its organization and manner of operation, together with further objects and advantages thereof, may best be understood by reference to the following description, taken in connection with the accompanying drawings in which:

FIG. 1 is a diagrammatic view showing a smoke detector being tested by the tester and method thereof incorporating the present invention;

FIG. 2 is a view of an aerosol container with a finger operated valve for selectively releasing the contents thereof to form an aerosol cloud;

FIG. 3 is a graph illustrating the effectiveness of the smoke detector tester in terms of time and distance.

FIG. 4 is a graph illustrating the particle sizes and relative concentrations of various types of smoke.

DESCRIPTION OF PREFERRED EMBODIMENT

Referring to FIG. 1, a smoke detector is illustrated in general by number 10 and is shown mounted on a ceiling 11 by any suitable means. The smoke detector 10 includes a vent or grill 12 through which the smoke or other products of combustion enter the detector. For testing purposes, a user 13 is shown with a hand held aerosol container 14 having a directional means for releasing the contents thereof into a sensing area or zone adjacent to detector 10. The contents of the aerosol can or container 14 simulate the properties of smoke or other products of combustion so that the smoke de-

4

tector 10 can be activated by simulated fire conditions. In this manner, the detector unit is tested not only for operability of its electrical circuitry but for its actual smoke and/or products of combustion sensing capabilities.

The aerosol can 14 as shown in FIG. 2 is of metal construction and widely used for varieties of pressurized consumer aerosol products. It is equipped with a finger operated plastic dispensing valve 16 with a very small aperture. The valve when depressed releases through the aperture a dispersion of the can contents 15 which is suitable for the purposes hereof. The preferred contents of the can 14 are a mixture of three ingredients, (a) seventy percent a hyrocarbon propellant, usually composed of 50% butane and 50% propane, said propellant is also an active ingredient in that it furnishes the majority of smaller particle sizes (0.5 micrometer and under), (b) (5% isopropyll alcohol, which serves as a carrier medium and (c) 25% dioctyl phthalate, an ester of phthalic acid, which provides the larger particle sizes (0.1 micrometers and over). All ingredients have boiling points of over 160° C.

Referring now to FIG. 3, it can be seen that the spray from the container 14 is most effective within a range of 0 inches to 5 feet (although it causes the alarm to respond at somewhat greater distances, up to 10 feet, with a correspondingly greater time lapse). Thus, a person of average height can easily test a smoke detector regardless of whether it is located on a wall or on the ceiling without needing to use a platform. Under repeated field trials, the alarm is actuated in less than 1 second and not more than 3 seconds. During the invention's intensive testing, no damage was noted involving either the circuitry or the housing of the smoke detectors used in the field trials. Further, the equivalent of eight years of normal usage of the product had no effect upon the sensing apparatus of the various brands of smoke detectors employed. A review of OSHA Regulations (29-CPR-1910-1000) indicates that the chemical formulation poses no danger to humans or to the environment under normal usage.

The graph of FIG. 3 also illustrates that the closer the spray or cloud is to the chamber 15, the faster the reaction time. This feature can also be converted to density of cloud particles causing the desired effect upon the voltage of the detection apparatus.

Referring to FIG. 4, it can be seen that the invention's test aerosol closely approximates the particle diameters produced by various types of smoke. Curve 17 is for smoldering wood smoke; curve 18 for smoldering PVC, curve 19 for burning benzene smoke and curve 20 for smoke from flaming paper.

In actual practice, the user aims the aerosol spray container 14 at the aera of the smoke detector 10 containing the vent, or grill 12. The cloud of spray 15 from the container enters through the grill into the sensing apparatus inside the detector. The pressurized propellant provides momentum for the particles to carry from the spray valve or nozzle 16 to the smoke detector. The numerals 15 in both FIGS. 1 and 2 signify the particle cloud as it approaches the smoke detector 10 after leaving the container 14. A four ounce (contents) container size was used in tests. Smaller and larger sizes would perform similarly.

While particular embodiments of the present invention have been shown and described, it will be obvious to those skilled in the art that changes and modifications may be made without departing from this invention in

4,301,674

5

its broader aspects and, therefore, the aim in the appended claims is to cover all such changes and modifications as fall within the true spirit and scope of this invention.

What is claimed is:

1. In a tester for simulating the presence of products of combustion for activating a smoke detector comprising a hand held pressurized container having a finger operated pressure release valve with a very small aperture which allows for direct spraying, a quantity of material stored in said container which when released under pressure will form a moving cloud of particulate matter simulating products of combustion, and a propellant included in said material which aids driving said cloud material in the desired direction so as to impact said smoke detector,

The improvement consisting of:

said cloud of particulate matter

comprising particle sizes which produce a mean geometric diameter of approximately 0.7 micrometers to 1.2 micrometers, thereby providing an appropriate range of particulate size such as to activate alarms in both the ionization and photoelectric type of detectors at sensitivity levels which indicate the detector is functioning as intended.

2. A tester according to claim 1 in which said cloud material consists of approximately seventy percent hydrocarbon propellant, approximately five percent isopropyl alcohol and the remainder dioctyl phthalate.

6

3. In combination with a conventional smoke detector, a tester for simulating the presence of products of combustion for activating said smoke detector comprising:

a hand-held pressurized container having a finger operated release valve;

a quantity of cloud material stored in said container simulating products of combustion;

a propellant included in said cloud material for driving said cloud material in a desired direction so as to impinge said smoke detector;

said cloud material includes particle sizes within the range of 0.01 micrometers to 5.0 micrometers, with a median geometric diameter of 0.7 micrometers to 1.1 micrometers;

said cloud material consists of approximately seventy percent hydrocarbon propellant, approximately five percent isopropyl alcohol and the remainder dioctyl phthalate;

said smoke detector is of ionization type or photoelectric type and said cloud material being further characterized as being effective to increase the resistance of the radiation source output so as to produce a voltage drop therethrough in the case of the ionization type detector, or to create a light scattering effect in the case of the photoelectric type detector;

said cloud material is characterized as having the required dispersion of particle sizes which simulate the early stages of fire.

* * * * *

10

15

20

25

30

35

40

45

50

55

60

65

APPENDIX D

Patent US4301674 - Smoke detector tester

method for manufacture

US6741181 May 17, 2001 May 25, 2004   System for testing a duct smoke or other hazardous gas detector and method for use thereof

US6812834 Jan 17, 2002 Nov 2, 2004 The United States of America as represented by the Secretary of Transportation   Reference sample for generating smoky atmosphere

US7567928 Jan 12, 2007 Sep 15, 2009 HSI Fire & Safety Group, LLC Method and apparatus for testing detectors

US8205476 Dec 24, 2008 Jun 26, 2012   Smoke detector testing tool

USD275183 Dec 28, 1981 Aug 21, 1984   Smoke detector tester

Claims

1. In a tester for simulating the presence of products of combustion for activating a smoke detector comprising a hand held pressurized container having a finger operated pressure release valve with a very small aperture which allows for direct spraying, a quantity of material stored in said container which when released under pressure will form a moving cloud of particulate matter simulating products of combustion, and a propellant included in said material which aids driving said cloud material in the desired direction so as to impact said smoke detector.

The improvement consisting of:

said cloud of particulate matter comprising particle sizes which produce a mean geometric diameter of approximately 0.7 micrometers to 1.2 micrometers, thereby providing an appropriate range of particulate size such as to activate alarms in both the ionization and photoelectric type of detectors at sensitivity levels which indicate the detector is functioning as intended.

2. A tester according to claim 1 in which said cloud material consists of approximately seventy percent hydrocarbon propellant, approximately five percent isopropyl alcohol and the remainder dioctyl phthalate.

3. In combination with a conventional smoke detector, a tester for simulating the presence of products of combustion for activating said smoke detector comprising:

a hand-held pressurized container having a finger operated release valve;
a quantity of cloud material stored in said container simulating products of combustion;
a propellant included in said cloud material for driving said cloud material in a desired direction so as to impinge said smoke detector;
said cloud material includes particle sizes within the range of 0.01 micrometers to 5.0 micrometers, with a median geometric diameter of 0.7 micrometers to 1.1 micrometers;
said cloud material consists of approximately seventy percent hydrocarbon propellant, approximately five percent isopropyl alcohol and the remainder dioctyl phthalate;
said smoke detector is of ionization type or photoelectric type and said cloud material being further characterized as being effective to increase the resistance of the radiation source output so as to produce a voltage drop therethrough in the case of the ionization type detector, or to create a light scattering effect in the case of the photoelectric type detector;
said cloud material is characterized as having the required dispersion of particle sizes which simulate the early stages of fire.

Drawings

Drawings

27911594R00051

Made in the USA
Lexington, KY
29 November 2013